SCIENCE AND ESP

International Library of Philosophy and Scientific Method

EDITOR: TED HONDERICH
ADVISORY EDITOR: BERNARD WILLIAMS

A Catalogue of books already published in the
International Library of Philosophy and Scientific Method
will be found at the end of this volume

SCIENCE AND ESP

EDITED BY

J. R. SMYTHIES

NEW YORK
HUMANITIES PRESS

Published in the United States of America 1967
by Humanities Press Inc.
303 Park Avenue South
New York, N.Y. 10010

© Routledge & Kegan Paul Ltd 1967

Library of Congress Catalog Card Number 67-24521

Printed in Great Britain

CONTENTS

PREFACE   *page* vii

1 IS ESP POSSIBLE?   1
   J. R. Smythies, M.D.
   Senior Lecturer, Department of Psychiatry,
   University of Edinburgh

2 PRESIDENTIAL ADDRESS DELIVERED
   AT A GENERAL MEETING OF THE
   S.P.R. ON 21 MAY 1952   15
   Gilbert Murray, O.M. (1866–1957)
   Regius Professor of Greek, Oxford, 1908–36

3 PSYCHICAL RESEARCH AND HUMAN
   PERSONALITY   33
   H. H. Price, F.B.A.
   Wykeham Professor of Logic, Oxford, 1935–59

4 NOTES ON CHANGING MENTAL
   CLIMATES AND RESEARCH INTO ESP   47
   Rosalind Heywood
   Member of Council, Society for Psychical Research

5 PSYCHOLOGY AND PARAPSYCHOLOGY   61
   Sir Cyril Burtt, F.B.A.
   Emeritus Professor of Psychology,
   University College, London

6 BIOLOGY AND ESP   143
   Sir Alister Hardy, F.R.S.
   Emeritus Professor of Zoology, Oxford

CONTENTS

7 THE NOTION OF 'PRECOGNITION'   page 165
C. D. Broad, F.B.A.
Knightbridge Professor of Moral Philosophy,
Cambridge, 1933–53

8 THE EXPLANATION OF ESP   197
C. W. K. Mundle, M.A.
Professor of Philosophy,
University College of North Wales

9 ESP IN THE FRAMEWORK OF
MODERN SCIENCE   209
Henry Margenau
Professor of Physics, Yale University

10 THE FEASIBILITY OF A PHYSICAL
THEORY OF ESP   225
Adrian Dobbs, M.A.
Member of Council, Society for Psychical Research

11 PSYCHOANALYSIS AND PARA-
PSYCHOLOGY   255
Emilio Servadio, M.D.
Psychoanalyst, Rome

12 C. G. JUNG AND PARAPSYCHOLOGY   263
Aniela Jaffé
Secretary, C. G. Jung Institute, Zürich, 1947–55

13 ANTHROPOLOGY AND ESP   281
Francis Huxley, B.A.
Research Fellow in Social Anthropology,
St. Catharine's College, Oxford

APPENDIX: A GUIDE TO THE EXPERI-
MENTAL EVIDENCE FOR ESP   303
John Beloff, Ph.D.
Lecturer, Department of Psychology,
University of Edinburgh

PREFACE

The problem of extra-sensory perception (ESP) is engaging increasing interest in scientific circles, both in America and the Soviet Union, as well as Western Europe. The experiments carried out over the last decades have accumulated a formidable array of evidence that ESP actually takes place. Since these phenomena are at first sight inexplicable in the context of contemporary science this gives rise to the problem: How are we to account for them? A few die-hard Newtonian mechanists claim that the scientists concerned have all been guilty of deliberate fraud. Other scientists are convinced by the evidence and claim that ESP has already been established. Others again would not go this far but would hold that a *prima facie* case had been made out for its validity and that what we should do now is to construct hypotheses to account for the phenomena and then test these hypotheses. What is apparent is that fewer and fewer scientists are merely uninterested.

This book presents the viewpoints of a number of people who have studied this subject. They include three philosophers, two psychologists, an anthropologist, a physicist, a psychoanalyst, a psychiatrist, a biologist, two parapsychologists, and a paper by Gilbert Murray read some years before his death in 1957. They discuss many aspects of the problem, but the general theme is that these phenomena are very probably valid, that they are important and that we simply cannot continue to ignore them. Several hypotheses are advanced to account for these phenomena that in principle are capable of further deductive development by those professionally equipped to do so (physicists, cosmologists, topologists, etc.) and which may eventually be tested by experiment.

I am very grateful to the contributors for their unstinting efforts: also to the Editor of the *Hibbert Journal* for permission to republish

PREFACE

Professor Price's paper, to the Editor of the *Proceedings of the Society for Psychical Research* for permission to republish Gilbert Murray's paper, to Mrs. Eileen Garrett for permission to republish Professor Mundle's paper and to Professor Gardner Murphy for permission to republish Professor Margenau's paper.

The Editor would also like to make it plain that his task in this has merely been of acting as the technical editor for the essays. The book itself was the idea of Rosalind Heywood and she did all the real work. She has for long been one of the mainstays of serious psychical research in this country and this volume is a fitting tribute to her inexhaustible devotion to the cause of scientific research into ESP.

I

'IS ESP POSSIBLE?'

J. R. Smythies

It is the province of natural science to investigate all phenomena in nature impartially and without prejudice. Among these natural phenomena one finds widespread beliefs among nearly all so-called 'primitive' cultures, and in many segments of advanced cultures, concerning such alleged human faculties as telepathy, precognition, divination, etc., and concerning such alleged facets of the human personality as the soul. The study of such beliefs is incumbent on anthropologists and their evaluation of such beliefs lies in the fields of philosophy and psychology. During the last eighty years there has been a good deal of research into these so-called 'paranormal' events. Many spontaneous cases have been studied and scientists have developed experimental methods of investigation. Some of this work has been poorly controlled and executed, but some has been very carefully carried out. Evaluation of the alleged spontaneous cases is extremely difficult as Professor Broad points out in his essay, because we cannot arrive at any estimation of the probabilities concerned. There is no way of distinguishing in each individual case between coincidence and some alleged paranormal manifestation. In the experimental work the probabilities are controlled. The results of all this work, it is generally agreed by even the severest critics, have led us in 1966 to the following position. We must *either* accept the validity of these phenomena *or* hold that all the workers reporting positive results (in experiments that stand up to the severest procedural analysis) are guilty of deliberate and often extremely ingenious and collective fraud. Professor Hansel, one of parapsychology's most energetic critics, has made it plain that these are the only alternatives before us. The experiments that have been carried out prove the existence of ESP beyond all doubt if, and only if, the experimenters did not deliberately falsify their results. Let us take an example. Suppose some scientists are conducting an experiment in telepathy. Scientist Brown prepares a pack of cards each with one of five symbols on it (cross, square, circle, star, wavy lines) in an order determined by a series of random numbers. This is handed to the agent under circumstances that preclude whispering of information. The agent sits by himself in a room, looks at each card in turn every five seconds and presses a button when he does so. He enters the symbols on a specially prepared sheet in duplicate, puts them in two envelopes at the end of the run, seals one with sealing wax, etc., and hands them to Brown. In another room the percipient is sitting with scientist Smith. The button in the agent's room is connected to a light in his. Every time this flashes he writes down a symbol on his sheet. At the end of the run he prepares one copy which he seals, and one copy which he hands to Smith. The two sealed copies are sent to scientist Jones. He verifies that the seals have not been tampered with and then scores the guesses against the targets. Finally Brown, Smith, and Jones meet to cross-check that Jones has not altered his copies since they must be the same as the copies held by Brown and Smith. Clearly this design will eliminate such sources of error as unconscious whispering, fudging of results by one experimenter, etc., but, equally clearly, if Brown, Smith, and Jones are in collusion to cheat, then they will be able to find means of doing so. Further precautions can be taken, other scientists can be recruited as further observers, etc., but again the claim that all these also enter into the conspiracy cannot be refuted, however fantastic the claim might appear.

However, a psychiatrist should be the last person to claim that otherwise honest and reputable people may not, under certain specific circumstances, enter into such fraudulent conspiracies if motivated to do so. Such motivations as a burning desire to save civilization from collapse by demonstrating the 'spiritual' nature of man, or desire for simple notoriety or some obscure unconscious urge could well induce such behaviour. This might indeed have happened in some instances. But it does not seem plausible to imagine that it has occurred to the extent that would be necessary to explain away all 'good' positive results on this basis. Hard-headed common-sense rebels. Yet, however *improbable* it is by no means *impossible*. Now Professor Hansel's case rests on Sherlock Holmes' dictum that when we have eliminated the *impossible* whatever remains, *however improbable*, must be the case. And Professor Hansel holds that ESP is *impossible* because it contravenes the order of nature discovered by modern science. Hence all claims to demonstrate ESP *must* be fraudulent.

What then, we must ask, is the basis for this claim that modern science has ruled ESP out of court? It runs as follows: The problem of ESP devolves on the nature of the human organism and the channels open to it for receiving information from the environment including other human organisms. Modern physics, it is argued, now presents a complete account, or at any rate practically a complete account, of the types of information channel usable by human organisms (light, heat, sound, etc.), and none of these could remotely be used in alleged ESP. Now it is still *logically* possible that event A in the environment could induce change A′ in the brain without any intervening causal chain. But this type of causal connection has never been detected under any other circumstances and it contravenes the principle of Uniformity of Nature. The brain is manifestly just a very complex collection of chemicals. It must therefore obey the ordinary laws of physics and chemistry. The *only* way in which the brain differs from any other part of the material world is its complexity. The human brain is by far the most complex structure, as far as we know, in the universe. Now it is possible that, just as different 'laws of nature' are operative when we are dealing with very large-scale events in cosmological astronomy and with very small-scale events in the atomic nucleus, similar changes could become necessary when we are dealing with the very complex. But none of the changes in the 'laws of nature' utilized by cosmologists and nuclear physicists tamper with a basic limiting principle so basic as the necessity of causal chains for connecting event A in one place and event B in another place, or if not a causal chain at least something like a gravitational field, or warping of space-time, etc. To avoid this gap in the causal chain one could postulate a 'complexity field'. That is to say that, whenever physical events are conjoined in a dense accumulation of extreme and diverse complexity such as the brain, then a new 'field' is set up by means of which information can be transmitted using this field. In that case a computer approaching the human brain in complexity would also be expected to demonstrate ESP. 'Complexity' here could be defined in terms of bits of information the machine could deal with per unit space in unit time. Physics, it may be argued, has not discovered this natural phenomenon because it has never examined systems of the required degree of complexity. There seems to be no logical reason why 'mass' or the 'electrical' properties of bodies should be the only properties that could give rise to fields. A 'complexity field' is even empirically defensible if predictions made from it were confirmed by experiment: for example, if the ESP capacity of a computer was found by experiment to be a function of its complexity.[1]

It is, of course, very likely that 'complexity fields' do not exist. But the fact that they logically could exist would appear to throw doubt on the claim made by Professor Hansel and other critics that ESP is *impossible*. Their claim is, however, that ESP is impossible within the framework of *existing* physics. Even this claim can be disputed as Mr. Dobbs does so ingeniously in his essay in this volume, using the sophisticated concepts of modern quantum theory rather than the ideas of nineteenth century physics on which the critics of ESP seem to base their objections.

Then again, the claim could be made that we know so much about the physiology of the human organism and the factors that control behaviour that we can exclude processes like ESP because modern physiology and psychology have shown that no such events occur. However, this claim is patently false. We know very little about how the brain actually operates and only some of the factors that control human behaviour. But, even if we knew exactly how the brain does operate in every detail of its biochemical and cybernetic mechanisms, this knowledge would necessarily have been obtained by the current methods of physiological research in which *all* parameters are controlled *except* the variable in which we are interested. The mere knowledge, then, of how the brain functions in conditions under which ESP was *not* operating is not any kind of evidence of how the brain could function under conditions under which ESP might be operating. Whether or not the brain—or the human being—has any ESP capacity can only be tested by experiments designed to do just this. The brain is

[1] This concept is akin to Marshall's suggestion presented in his paper 'ESP and memories: a physical theory', *Brit. J. Phil. Sci.*, Vol. X, 1959–60.

clearly no more than a very complex physico-chemical machine. But the current methods of physiology by themselves will not tell us whether it is a machine whose function is merely to shuttle nerve impulses about (plus electrotonic potentials, D.C. potentials, etc.), or whether in addition it is a machine designed to pick up minute 'mind influences' as Eccles has suggested. Either is equally logically possible.

Furthermore, the claim that we can explain all human behaviour in purely behaviouristic terms remains at present a pious hope; or rather a working hypothesis for a certain programme of research in psychology. Much of human behaviour can certainly be explained in terms of brain mechanisms or in terms of ordinary environmental influences. But the fact that it may logically be *possible* at some far distant date to explain all human behaviour in these terms does not in any way allow us to deduce that this need ever be the case. And, as we have seen, any claim ever to have done this will be invalid *unless* specific experiments to test for the occurrence of ESP (carried out intelligently and without prejudice) were included in our behavioural analysis. Professor Hansel has mistaken the blueprints of a working programme in psychology for some immutable 'natural law'. It is hardly reasonable to suppose that man, after some 400 years of scientific endeavour set in a universe with a time span of some 4,000,000,000 years, can have discovered all that there is to know about reality. I doubt if any physicist would be so presumptuous as to make this claim. Therefore it might seem preferable to set ourselves more modest aims. It is possible that ESP is a fact. Can we, then, construct any theory to account for it? The real objection of the neurophysiologist to dualistic theories of mind have been their untidiness. He can present a coherent account of the brain's operation in terms of the firing of nerve cells under the influence of excitatory and inhibitory synapses whose chemical mechanisms are rapidly becoming understood. It is irritating to be asked to add to this coherent and consistent scheme the alleged activity of 'mind influences' about whose properties or nature no suggestions of any kind have been made. Unless specific proposals can be made as to the nature of these 'mind-influences' and to the nature of their interaction with events in the brain, we are hardly likely to make much progress.

We have seen that the claim that ESP is impossible because there is no place for it in contemporary science is logically invalid.

Whether ESP occurs or not is a matter of fact. The best way to proceed, therefore, seems to be to construct as many hypotheses as we can as to its mode of action, to subject these to deductive development, and then to try and decide between them by experiment. How, then, could we explain ESP? This may perhaps be more accurately phrased as follows. How could a human organism obtain information about the environment ('clairvoyance') or from another human organism ('telepathy') without the use of any of the recognized channels of sense? The following methods are *possible*:

(i) By some form of electro-magnetic (radio) transmission between brains. For reasons detailed in this book this does not seem to be very likely.

(ii) By some as yet undetected physical 'force' such as the possible 'complexity field' suggested above or by the types of operation discussed by Mr. Dobbs in this volume. These explanations are open to the objection that the operations they describe would appear to have no clear-cut *function*.

(iii) The most plausible explanation, to my mind, is provided by non-Cartesian dualism. Cartesian dualism stated that the world consists of the physical universe extended in space and a number of individual minds not extended in space. Non-Cartesian dualism suggests that the world consists of the physical universe extended in physical space and a number of substantive minds extended each in a space of its own. The totality of each individual consciousness (composed of sense-data, images, thoughts, and the Ego) is located in its own space-time system, a *different* space from that of the physical world. This theory has recently been elaborated at length by Professor H. H. Price and myself (Smythies, 1966). Two different versions of it are current. In Professor Price's original theory there are no spatial relations between mental space and physical space. The relations between them, or rather between the events contained in them, are *causal* and *temporal*, but not *spatial*. Nevertheless, there is a constant flow of information between the two since the sense-data that represent the external world are on one side of the barrier—the 'mind' side—and the physical world itself is on the other. This information, it is postulated, is carried by a non-spatial information channel.

The other version of the multiple space theory was originally put forward by Professor Broad. In this we postulate one four-dimensional physical space-time as well as many individual 'mental' space-times (each containing the sense-data, images, thoughts and Ego of an individual). Together these form one single n-dimensional space-time continuum. One *cross-section* of this *is* physical space-time and other cross-sections *are* mental space-times. Thus when we look at a man's brain and ask, 'Where is his (substantive) mind? (his sense-data, images, Ego, etc?),' the answer may be 'In another space higher-dimensional relative to the space in which his brain is'. The physiologist may ask, 'If you say that the mind is outside the brain, what on earth do you mean by outside? How could "influences" from such a mind interfere with the chemical and electrical mechanisms of the brain?' He may find it impossible to envisage any 'psychic factors' from a non-spatial mind altering the fine detail of synaptic events in the brain in apparent defiance of the laws of physics and chemistry. One source of this difficulty is that it is impossible for us to imagine a non-spatial entity—we cannot form a mental image of it—and *this* is because mental images themselves are *spatial* entities—which is one of the planks of the non-Cartesian criticism of Cartesian ideas of the criteria for distinguishing the mental from the non-mental.

However, the alleged interaction between mind and brain becomes possible to envisage if we postulate that the total human organism is extended in an n-dimensional space (where n may be 5, 6, or 7 depending on the particular geometry involved). One section of this space contains his physical body and brain. The other contains his sense-data, images, etc. (see figure 1). The two may be linked by channels of information that can be depicted in the diagram by a series of vectors at right-angles to all vectors used to depict interactions in the brain itself. The geometry of such a system allows for the closest contact between a substantial mind and its brain. In this case, when the physiologist asks, 'Whence comes the influence that is alleged to act on events in the brain?', the answer may be 'From another space which forms together with the brain space an n-dimensional manifold'. It is remarkable that so few people have questioned so arbitrary an assumption that the universe of events is limited to one single four-dimensional space-time system.

A physics based on this assumption can give a perfectly adequate picture of the universe *except* (i) for events in the other parts of the universe—the 'mental' parts, and (ii) for interactions between these two. Until recently all the events in the various mental worlds have been denied independent existence under the joint

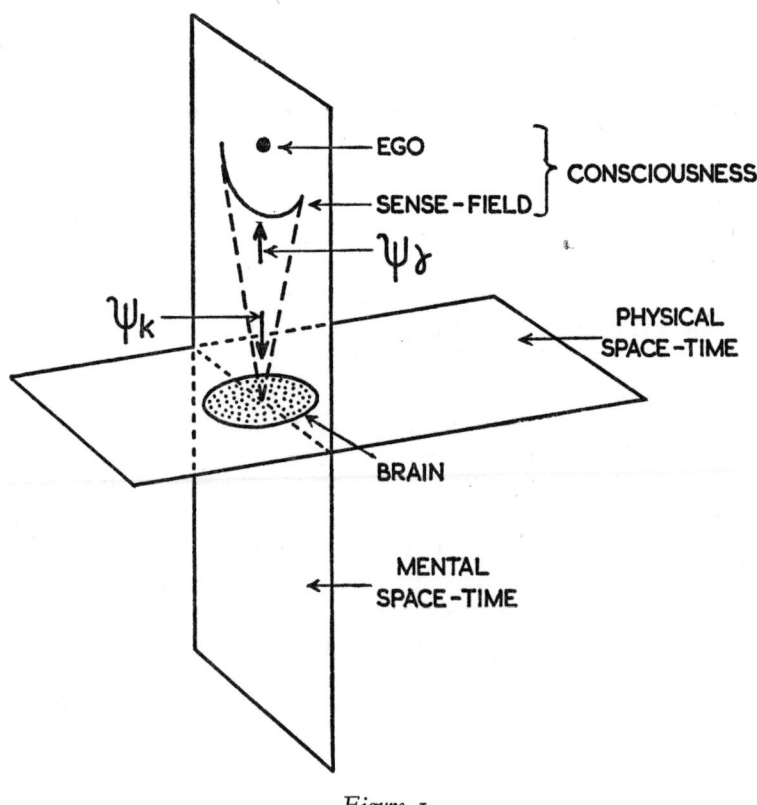

Figure 1

influence of the theory of psycho-neural identity and the theory of naïve realism: this entailed that no interaction between the two realms was logically possible. An n-dimensional universe of the kind postulated above would be in no way 'immaterial' or 'idealistic'. Every principle of scientific materialism remains unaltered. The universe is still a manifold of *events* interacting causally in a space-time manifold. The only difference is that the space-time manifold has a more complex geometry and there are more events than has hitherto commonly been supposed. No new principles are involved. Certainly the postulate that real events in the universe are ordered in an n-space-time is child's play compared with some of the concepts currently accepted in quantum physics! (See Mr. Dobbs' paper.) The reasons for suggesting such an hypothesis are as follows: (i) As outlined in *Brain and Mind* it enables us to construct a very much simpler theory of perception than any extant; (ii) it offers possibilities for deductive development by topologists, geometers, and physicists; (iii) it offers a simple explanation for ESP; and (iv) it is possible that deductions may be made from it that may be testable by experiment.

The first reason has been discussed already in the book mentioned, and it is only relevant here in that it is advantageous for a theory, which *can* account for ESP, to carry with it fringe benefits in the form of improved theories in another field.

As for the second reason, the physics of interaction in a four-dimensional space-time continuum has been exhaustively developed. We can now ask what could be the possible physics of interactions in an n-dimensional continuum? Furthermore, there would appear to be many different ways of combining a number of different four-dimensional space-times into one common manifold. Consider, for example, a cube. This can contain a number of planes, and these may or may not intersect. Let us imagine two planes (one of which, A, represents the physical, and the other, B, a mental world) that fail to intersect. Therefore the inhabitants of Flatland (to recall E. A. Abbott's telling little fantasy of that name) confined to one plane would never be aware of events in the other plane. Let us now take a third plane, C, that intersects both A (the physical) and B (the mental), and let us suppose further that the inhabitants of Flatland are really dualistic beings. They have physical bodies in the first plane A, minds (including all their sensory fields in the second plane B) and that communication between the two is conducted by the third plane or communication channel that connects the two (figure 2). Now if events in B are so arranged to represent events in A (in the manner described by representative theories of perception) it will seem to the naïve Flatlander that he is totally a being moving around in A. The whole of B and C will be out of reach of his sense-organs in the most literal sense. He will have no inkling of their existence. If he takes a naïvely realistic attitude towards his own sensory fields he will assume that these give him a 'direct view' of the physical world in A rather than a mediated or television-like view conducted by a mechanical system from his 'brain' in A to his sensory field in B via the communication channels in C. This is only a crude picture attempting to illustrate in a three-dimensional world what may in reality take place in a six-dimensional space. But, since it is difficult to visualize events interacting in n-dimensional space,

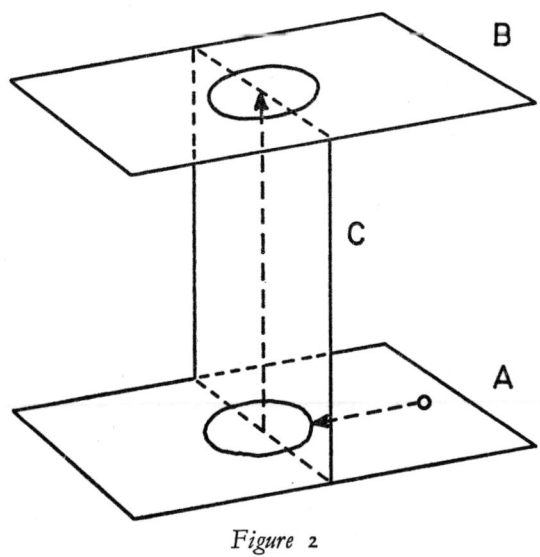

Figure 2

such a model, although very rough and ready, may make it easier to grasp something of the logic of these ideas. Take again the example of our Flatlander living in his plane. He may have no conception of the vast expanse of the cube that we three-dimensional beings can easily see surrounds him so completely. Similarly, a brain to us appears to have but one avenue of approach—through the skull. But every part of it will be immediately accessible from any extra dimensions of space that may be juxtaposed. If causal influences can cross the 'interface' between one of these systems and another it is possible that they may set up peculiar effects in the target area that might be detectable by instruments located in that area. To leave Flatland and return to the real world, it is conceivable that a topological analysis of the properties

IS ESP POSSIBLE?

of the different possible arrangements of more than one 4-space-time into common manifolds, plus a physical analysis of what causal interactions between these systems might be like (looked at from one such system alone and without access to the other nor to the interface between them) may lead to predictions that could be verified by experiment. Just as we can see how the unwitting Flatlander is related to the enveloping cube, so a being able to apprehend events taking place in an n-dimensional space might be able, at one glance, to 'see', as it were, how we 'cube-landers' are related to the n-spaces that may surround us in turn. But since we cannot do this and are thus in the position of the poor Flatlander trying to cope with his baffling and unfathomable cube, we must depend on more subtle forms of analysis, if such are possible.

The following is a possible hypothesis as regards the relation of non-Cartesian dualism and ESP. Non-Cartesian dualism suggests that one function of the brain is to collect in a highly 'poised' form (to use Eccles' phrase) a processed model of external reality within itself. Between this model and the final end-product—the sensory fields in consciousness—some channel of communication must exist. This is a specifiable form of the $\psi\gamma$ process postulated by Thouless and Wiesner (1947).[2] As a rule the input to this channel is via the brain, but on occasion it can be directly from some external object, including someone else's brain. The sense-organs of the physical body, the brain, $\psi\gamma$[3] and the sensory fields in consciousness may be considered to be one single *mechanism* for the transmission of information—logically a representative mechanism no different from television except that it operates in n-space-time rather than 4-space-time. The 'focus' of $\psi\gamma$ is normally on the brain, but it might well have a 'penumbra' capable of being influenced directly, say, by a pack of Zener cards. Owing to the n-dimensional geometry of the system, the distance between the percipient and the target in physical space might very well not have any effect on $\psi\gamma$, in which case ESP would not be subject to any reduction with distance as the experiments conducted so far suggest to be the case.

[2] R. H. Thouless and B. P. Wiesner, 'The Psi process in normal and paranormal psychology,' *Proceedings S.P.R.*, Vol. 48, pp. 177–96, 1947.

[3] Here taken as a symbol for the information channel linking brain and consciousness on the input side. $\psi\kappa$ is a symbol to represent the possible output side of this channel.

Lastly it might eventually be possible to detect by means of physical instruments influences entering the physical universe from outside it—using outside in the sense that I have described. At any rate, since a number of theories to account for the existence of ESP are logically possible—they describe logically quite possible universes—they might dissuade philosophers from trying to prove that all dualistic theories are impossible on logical grounds and psychologists from holding to the opinion that ESP is impossible.

This theory can also be modified to account for precognition as follows. We can postulate that the physical world consists, not of an array of three-dimensional objects existing in time, but rather of four-dimensional array of objects (or 'world-lines' in the terminology of relativity) *along* which our field of observation is travelling. This is a Hinton universe and its relation to precognition has already been discussed by J. W. Dunne[4] and by Professor Broad.[5] In the previous versions, however, the nature of this 'observer' was never clearly specified except in a vague Cartesian sense. The general principle is clear enough. If we draw a line on a piece of paper and move a narrow slit along it from *A* to *B*, the one-dimensional line takes on the appearance of a dot moving about itself—when what really is moving is the 'field of observation'. Hinton and Dunne merely applied this argument to real objects in the real world. The theory of non-Cartesian dualism can be adapted to this model as follows. We may suppose that 'mental' space (or rather the consciousness contained therein) is in relative motion to physical space in which real objects are four-dimensional solids and not three-dimensional ones. Figure 3 shows in the form of a crude diagram how this system operates. *ABCD* represents the physical world in which the world-line of the observer's brain is drawn. *AB* represents the time axis and *AD* represents all three space axes. Objects thus literally stretch into the future and into the past as frozen four-dimensional solids. In other words, ancient Rome is still 'there' together with the physical bodies of the ancient Romans in the form of four-dimensional solids past which the 'field of observation' swept two millennia ago (*x* kilometres in the *AB* dimension). *EFGH* represents the

[4] J. W. Dunne, *An Experiment with Time* (London: Faber, 1939).
[5] C. D. Broad in *Religion, Philosophy and Psychical Research* (London: Routledge, 1953).

IS ESP POSSIBLE?

mental space of one individual where the observer's sense-fields and Ego are located. *IJ* represents the volume of intersection of the physical and psychical worlds across which the channel of information flow passes from the brain in *ABCD* to the sense-fields in *EFGH*. *EFGH* is sweeping along *ABCD* (from *A* to *B*).

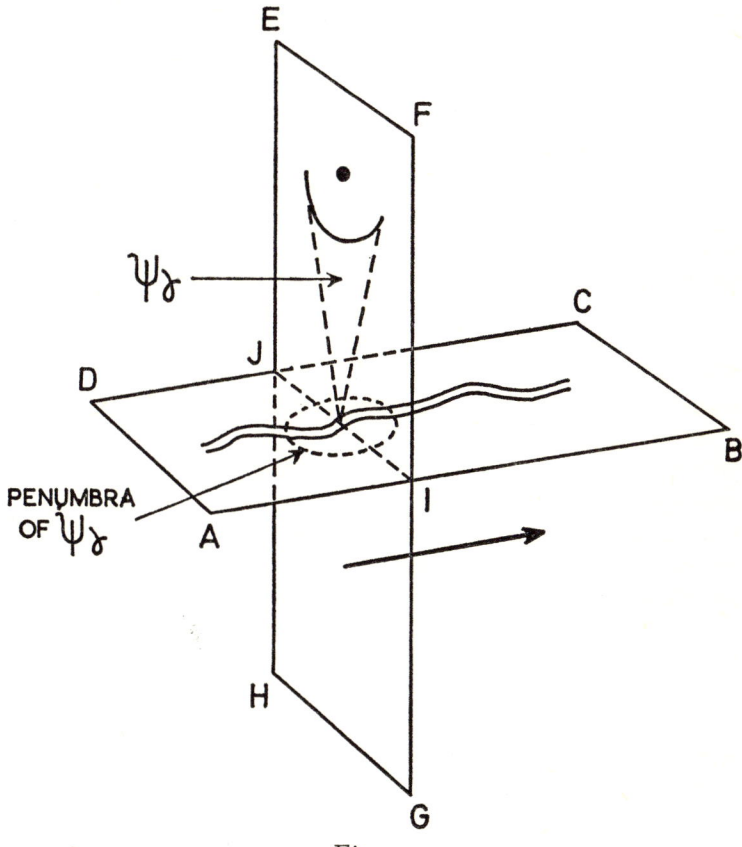

Figure 3

In the previous statement of non-Cartesian dualism we operated with the orthodox concept of time. Physical events and mental events, although located in different spaces, shared the same time. In this alternative model we need two times. 'Orthodox' time has been spatialized and forms the *AB* axis. 'Real' time is that in which *ABCD* and *EFGH* are in relative motion. This relative motion *generates* ordinary (subjective) time, as well as the 'movement' of objects in the world, from their real frozen four-dimensional shapes. This is the familiar Dunne–Hinton model. Dunne made the mistake, however, in assuming that this progression from T_1 to T_2 necessarily involves an infinite regress. Broad[6] has subsequently shown that it does not. This is a perfectly possible and valid theory of the Universe and the nature of time. To account for precognition all we have to suppose is that the 'penumbra' of $\psi\gamma$ (that we supposed might extend into the physical world outside the brain to account for telepathy and clairvoyance) extends also along the AB axis in figure 3—i.e. into the 'future'. The Hinton universe does not necessarily imply a gloomy Calvinistic determinism as has been supposed. The *shape* of objects in *IJBC* (that will *become* events when *IJ* sweeps past them) may be modifiable by events in *IJ* (i.e. by the exercise of free-will in the present): that is the world-lines may be more like ropes than steel rails. Thus *IJBC* would contain only the *most probable* future, somewhat in the manner that Mr. Dobbs describes for his system of two times. Although it must be noted that his two times are quite different to the two times postulated by Dunne. His two are a 'real' time and an 'imaginary' time (as specified in detail in his essay in this volume). The two times specified here are, respectively, spatialized time (T_1) and real time (T_2). The Hinton universe is certainly a very interesting one that has been somewhat neglected in recent years. As far as I know only Professor Broad has discussed it and has pointed out some of its remarkable advantages—for example its parsimony; since all movement in the universe and hence, all laws to explain them, is reduced to the physical geometry of four-dimensional objects and the movement past these of the field of observation. 'The shape of things to come' may have a more literal meaning than Wells ever supposed.

[6] Broad, *loc. cit.*

II

PRESIDENTIAL ADDRESS*

Gilbert Murray

Delivered at a General Meeting of the Society on 21 May 1952

I observe that almost all the most impressive addresses given of late to this Society have begun with an apology. They come from highly qualified philosophers, like Professors Broad and Price, who can analyse the implications of the evidence, or at least of the current interpretations of it; or from long and exact students of the history of the S.P.R., like Mr. and Mrs. Salter; or from experts who have observed or conducted long series of scientifically controlled experiments. How much more humbly must a complete non-expert like me make his apologies, when attempting to discuss the whole field of Psychical Research, and in part to explain the one corner of it in which he has had some personal experience.

On his last visit to Oxford William James once said to me, in a discussion about the future of religious belief, that he thought it would be largely affected by the result of the researches of this Society. The statement made me reflect. If we take religion in a narrow doctrinal sense, there has certainly been a great liberalizing and internationalizing movement. The Pope has lately been addressing a collection of twenty-four thousand boys and girls, some Protestant and some Catholic. On committees at Geneva or Lake Success, Moslems, Jews, Hindus, Christians, and Buddhists have worked together for various beneficent purposes without feeling any call to address each other as Unbelieving Dogs. But I do not think this Society played any part in that great change. If, however, we take 'religion' in a very wide sense, as meaning

* Reprinted from *Proceedings of the Society for Psychical Research*, Vol. XLIX, Pt. 181, Nov. 1952.

what has been called the 'inherited conglomerate' of beliefs, habits, expectations, approvals, and disapprovals dominant in a given society at a given time, I think one's judgement would be rather different. There have been during the last two or three generations in England some large and surprising changes of outlook, in some of which I should judge that the S.P.R. has played a rather interesting part. We must not exaggerate. There is much truth in Andrew Bradley's statement that the supposed Victorian family is the greatest work of creative imagination that the twentieth century has produced. Nor must we forget that the 'inherited conglomerate' is always to some degree in a flux, varying from old to young, from generation to generation, and of course varying widely in the same generation between the educated and uneducated. Still the changes of the last seventy or eighty years have been rather exceptionally marked.

Of course, the political, social and economic problems have greatly changed. But I doubt if Psychical Research has had any great effect on them. In science itself the advances have been almost revolutionary, especially, I suppose, in physics. I was brought up to believe that the types of certain truth were Euclid's theorems and Newton's law of gravitation. Einstein showed them both to be inadequate and in a sense unreal. What was one to believe after that? I was brought up to believe that the Earth was gradually cooling and that, like the Moon, she would become too cold to support life. Then, on the contrary, I learned that she would become intolerably hot; and later, that neither view was true, she would explode. I was told that the sun was a fixed star. Not at all; the whole universe was receding at enormous speed; that it was expanding; that it was coming into existence; and lastly, when the layman's mind was already reeling, that space itself was curved—whatever that might mean. The orthodox conglomerate was wonderfully open-minded and ready to welcome new ideas, except indeed just in the region that interests us most. There it was decidedly intolerant, would stand no nonsense. It would not listen. A belief in hypnosis, for instance, now a well-ascertained fact, was beyond the pale. The fact that various strange phenomena, which we now explain as hypnotic, were handed down in the popular tradition, told against them rather than for them to the scientific convention of the day. They seemed to be only old superstitions revived. The Viennese physician Mesmer, whose hypnotic cures had spread his fame all over Europe, was treated as a charlatan, examined in a hostile spirit and finally discredited. A generation later Elliotson, Professor of Medicine at London University, was ordered by the authorities to discontinue his hypnotic experiments, whereupon he resigned his chair. About the same time Esdaile, a surgeon in the Indian Service, performed some three hundred operations under hypnotic anaesthesia, but medical journals refused to publish his reports. In 1842 W. S. Ward in London amputated a thigh with the patient under 'mesmeric trance', and reported the case to the Royal Medical and Chirurgical Society. The Society heard, but refused to listen; the patient was accused of being an imposter, and the record of any such paper having been read was struck from the minutes of the Society. It was not until nearly a hundred years after Mesmer's chief cases, when Charcot took up hypnosis at the Salpetrière, that hypnosis gained full credence and was accepted as a branch of medical practice.[1] A very similar history of intolerance could be told of faith-healing, telepathy, and various other phenomena. A great part of the whole science of psychology has developed from the systematic observation of phenomena which would have been scornfully set aside a hundred years ago as superstitions unworthy of attention.

Do not let us be unjust to this sceptical or negative attitude. We must not forget how close Europe still was to very cruel and revolting superstitions. England itself was comparatively safe. The statute prescribing the burning of witches had been repealed in 1738, though Ruth Osborn was ducked as a witch and then murdered by a mob in Hertfordshire, close to London, in 1751. But in Ireland, in my own lifetime, a child, who was for some reason reputed to be a changeling, was beaten and burned with irons, the mother being locked out of the room while the invading fairy was exorcized, though unfortunately the child died in the process. A witch was burned, in a village near Monte Cassino, so I am told by a friend who lived there, in 1912 or 1913. I knew an Englishman who somehow lost his memory in Italy, and was found some days after tied up in a market-place as a madman and beaten to drive out the evil spirit. In northern Greece a friend of mine found a madman tied up in a public place for anyone passing to

[1] I have quoted the above almost verbally from M. Polanyi's *The Logic of Liberty* (London: Routledge, 1951).

beat as he chose. Such possibilities were near enough to have left behind them a real horror. So no wonder the men of science felt that the great need of the time was not to search sympathetically for such elements in old beliefs as might be really true, but firmly to reject the whole mass of degrading and inhuman nonsense.

None the less this excessively sceptical attitude provoked some reaction. There really are more things in the world than the science of any period can fully account for. One can see the feeling of this in many leading nineteenth-century thinkers, such as Carlyle and Ruskin, and even in J. S. Mill himself. And, of course, a much grosser kind of superstition makes a lasting appeal to human nature. The preacher who cries in Carlyle's words, 'Come unto me ye who hunger and thirst to be bamboozled,' will always find a response. Among the remote and uneducated, particularly perhaps among the Celtic fringes, there was generally a survival, or sometimes a passionate resurgence, of unauthorized supernatural beliefs.

Much of the boldest and most uncritical offensive, however, came from America. The fact is patent, and I think one can see the explanation. In that great democracy the common man, however unqualified, has pretty full freedom of speech. The aristocratic tradition in England and Europe generally makes the uneducated rather timid about asserting their own ideas in public against those of the experts. The actual founder of American spiritualism is said to have been Andrew Jackson Davis of Poughkeepsie, who was thrown into genuine trances and expounded a mystical doctrine of spirit communion, called Harmonial Philosophy, in several large volumes.[2] But Spiritualism as a vigorous popular movement seems to have started chiefly from the Fox sisters. The two younger produced communications from the spirits by means of raps, and also various poltergeist phenomena. Both, it seems, were remarkably attractive young women, while the mother and eldest sister were experts in salesmanship. They had astonishing success both in America and in England, where so great a scientist as Sir William Crookes was convinced by them. Their father, however, had been a drunkard, and the girls eventually took after him. Kate wrote and withdrew confessions proclaiming that the phenomena were all a fraud—excepting, curiously enough, the actual raps, which were in some sense genuine.

[2] F. Podmore's *Modern Spiritualism*, Vol. I, p. 158.

PRESIDENTIAL ADDRESS

Margaret confirmed the confession, and the whole business ended in singular squalor. One cannot but suspect that if the arts of observation and detection had been better developed in the middle of the last century, many of the wonder-workers of the time, such as the Foxes, Florence Cook, and even the great D. D. Home himself would, like Mme Blavatsky, have had more interrupted careers.

Certainly a disproportionate number of mediums and thaumaturges and founders of new religions were American. Not many European countries can produce figures like Joseph Smith or Brigham Young, the founders of Mormonism, or Mrs. Eddy, the founder of Christian Science, or even Dr. Buchman, not to speak of Amy Macpherson and hundreds of persons of equally vulnerable pretensions. One popular preacher in Chicago who used to send me literature discovered that he was the prophet Elijah reincarnate, and in order to convince possible doubters, took to wearing large wings. One could quote such extravagances by the dozen. But one strange development which could not, I think, have occurred except in America was the history of the New Motor which was to save mankind. A man with an enthusiastic faith in machinery, though no great knowledge of its working, felt it to be obvious that Salvation, like everything else, could be much more effectively produced by machine power than by human labour. Mesmeric, or as we should say hypnotic, influences were then called 'animal magnetism', and, since the earth was a great magnet, there was obviously a tremendous store of mesmeric force there ready to be tapped. By the guidance of a series of dreams this man succeeded in building a motor which was to concentrate in itself the magnetic force of the earth and so manufacture Salvation on a world-wide scale. With the help of subscriptions it was made and set up, but somehow would not work. Various expedients were tried. The Faithful gathered round it in prayer, but in vain. Presently a leading spiritualist was called in to advise. He decided that the dreams were genuine and came from the spirit world; but that any engineer would see that the machine could not move without breaking itself, and perhaps the spirits had wished to try or merely to tease the inventor. Meantime, however, a woman in the south, several hundred miles away, had it revealed to her that what the machine wanted was a mother, and that she was called upon to assume that sublime office. She came to see the inventor, who reports with obvious good faith: 'I did not quite understand what she wanted, but I gave her the key of the shed.' But even when mothered it would not move. At last someone suggested that it was in much too high a position. It was not near enough to the magnetic centre of the earth. So some hundred or so of the Faithful harnessed themselves to it and dragged it over miles of farmland to the bottom of a river valley. There, no doubt, it might have performed better, but the ignorant farmers of the neighbourhood broke it up and threw it into the river. The story is told in detail in Mr. Podmore's book, *The New Motor*. Fashions change, but the interest in the supernormal continues. A book published in 1946 reports that there were then twenty-five thousand practising astrologers in the United States.

We must not forget the English newspapers which employ a regular astrologer, nor yet the temporary vogue among more intellectual circles of Mme Blavatsky and her brand of theosophists. My point is to illustrate the great mass of utterly unacceptable material which lay before the scientists of the mid-nineteenth century. First there was a vast respectable tradition of miracles and wonders, much of it supported by religious doctrine; next, a large but indefinite remnant of primitive superstition among the uneducated; thirdly, the constantly recurring interest, I had almost said the craving, among educated and uneducated alike, to discover, or hope to have discovered, some certainty behind the veil. It is also worth remembering that, overwhelmingly strong as the tradition is of the existence of prophets and *shamans* with superhuman powers, it is always accompanied by a suspicion of possible fraud or false pretensions. And the same suspicion accompanies the modern evidence. Mrs. Salter records that in her childhood she saw something of the famous medium Eusapia Palladino, and though she did not attend any of the actual seances, remembers that in unprofessional moments Eusapia cheated at every game she played.

We can understand the indignant phrase of my old colleague Lord Kelvin, that all the phenomena were 'half fraud and half bad observation', as well as the wiser conclusion of Professor Sidgwick that, in the face of such a bewildering mass of remarkable and ill-attested phenomena, it was 'a sheer scandal' that they should be left with no serious attempt to find out what parts of them, if any,

PRESIDENTIAL ADDRESS

were true. That, of course, was the purpose with which this Society was founded, and the quest on which it has been engaged for over seventy years.

What have we actually discovered? It is hard to say. Hypnosis is now accepted as a *vera causa* in medical science. Much akin to hypnosis are various forms of psychotherapy recognized and practised in hospitals. Going a step farther, I think we are bound to admit the fact of actual faith-healing. The evidence from Lourdes and other Christian shrines is very strong, and is confirmed by similar or even more abundant evidence from Hindu shrines; nor should we forget the successes of Christian Science. The actual limits of faith-healing must be left for medical science to determine. The fact of immediate relief is certain; wounded men in great pain calling for morphia have fallen peacefully asleep on receiving an injection of pure water. Continuous relief in chronic cases seems certain, and by the relief of anxiety and a consequent lessening of the flow of blood to the affected part, has sometimes been hard to distinguish from actual cure. I have known one case of this in my own family. We need not therefore be quite as puzzled as the seventeenth-century Dean of Wells in Aubrey's *Miscellanies*, who writes: 'the curing of the King's evil by the touch of the King doeth puzzle my philosophie; for, whether they were of the house of York or of Lancaster, it did'. Similarly, when Freud released many patients from dangerous repressions by getting them to remember some forgotten incident of their childhood, in some cases it turned out that the incident had never really taken place. It was just imaginary. But the cure worked. Then, again, every anthropologist will remind us that Faith can kill as well as cure; there are well-proven cases from New Zealand and the Pacific Islands of people who have died because they believed they had been touched by a magician or highly tabu chief. My brother in Papua saw a man give another a handful of pebbles. 'What are these?' said the man. 'They are sent specially to you by so-and-so,' was the answer, mentioning the name of a well-known sorcerer. 'Oh, then I am done for,' said the victim, and died that night.

Of all the problems that faced the S.P.R. on its foundation, the first and greatest, I suppose, was the problem so confidently answered by the Spiritualists: do we in some sense survive our bodily death, and is there communication between the living and the spirits of the dead? The subject is too large and important to be treated in passing. We may note Mrs. Sidgwick's conclusion, reached after considerable study, that the evidence of survival did not amount to proof, but was enough to justify personal belief.

Members of the Society will remember the two attempts that were made to obtain a wide Census of Hallucinations: it came out in both that roughly 10—11 per cent of those questioned had had hallucinations, and of the hallucinations about 10 per cent appeared to be veridical. A striking attempt was made by Mr. Podmore to show that all such phantasms were phantasms of the living, not of the dead. This would apply to one striking case of which I had some knowledge, and which is in one point very remarkable. The phantasm of an intimate friend of mine appeared early one morning on 3 or 4 September 1898 to a lady he knew in London. She felt, as usual in such cases, no particular surprise, and reported that he was smiling and said, 'I am going on on the 11th.' It was just after the Battle of Omdurman. My friend survived the battle, but was killed the day after. The phrase 'on the 11th' had no ascertainable meaning; what he really did was to 'go on *with* the 11th', that is, the 11th Lancers, a regiment to which he did not belong. Must he not really have said, 'I am going on *with* the 11th', and been misheard? If so, it would seem to follow that the phantom was not merely a creation of the percipient's mind, but was carrying a real message. It is, of course, extremely difficult in many cases like these to prove either a positive or a negative. But my own impression is that most of the commonly reported wonders, both traditional and new-fangled, have so often been proved to be either misreported or misobserved or sometimes simply fraudulent, that they must be regarded, to say the least of it, with extreme suspicion; I would include in this category spirit photographs, haunted houses, extensions of the human body and the great majority of poltergeists.

A new standard of strictly scientific observation of these supernormal phenomena is evidently a great desideratum. And for one class of them such a standard has been successfully set by Dr. Soal in his statistically controlled experiments with Mr. Shackleton and Mrs. Gloria Stewart, and on a greater and more elaborate scale by Professor Rhine and his staff at the Duke University, North Carolina, where there is a special Institute of Parapsychology with fifteen rooms and a staff of six to eight whole-time parapsychologists.[3] I cannot criticize the work of the Institute, except to say that the good faith of the workers seems undoubted, and the accuracy of the methods well attested, yet the statistical results reported are—to me at least—quite incredible. In the field of Precognition especially there are parts as to which I find myself belonging to the class described by Dr. Soal as 'hopelessly prejudiced by some outmoded philosophy which they probably imbibed in their youth and which they are too old to abandon'. I have always held, in accord with all my scientific friends, and with that admittedly dangerous guide, Common Sense, that the cause which produces an effect must come before the effect; the cause precedes, the effect follows. I know, of course, that Time is called a Fourth Dimension; and I fully recognize that the exact placing of any event requires three dimensions in space and one in time. But surely there is nothing magical in that. It is not the sort of fourth dimension which would enable us, for instance, to see and touch the inside of a solid. I quite see that our whole conception of events in the Universe must be conditioned by the limitations imposed on our minds by our bodily structure; that the whole world would be exceedingly different to us if we could really

'   know, hear and say
What this tumultuous body now denies,
Feel, who have put our groping hands away,
And see, no longer blinded by our eyes.'

What repels me is the supposition that sometimes, some few of these limited minds should, for no ascertainable reason, completely overcome the limitations of human reason in one small point, while leaving all others unchanged. If I go with Alice into a Looking-glass world I shall expect to find that left is right, that people cry because they are going to be hurt, and pick themselves up because they are going to fall down. But I should at least expect some consistency. I look for some other explanation. I may add that I have read Mr. Dunne's book twice in the hope of being convinced, but have not been. I think that the reasoners who do magic with a fourth dimension are like those who draw conclusions about the real world from the use of surds in

[3] See the description by Dr. West in the *Journal S.P.R.*, XXXV, Pt. 656, January–February 1950.

mathematical formulae. I am, of course, ready to accept mathematical calculations which make use of the square root of -1, or the convention that A to the power of 0 equals unity, but I do not believe that the convention is more than a convention or that I shall ever meet the square root of -1 in real life. Consequently I feel enormous difficulty in accepting some of the statistical phenomena of precognition which, I confess, I am unable to explain otherwise. My old-fashioned mind notes with much comfort that Dr. Soal himself feels doubts about precognition by pure clairvoyance with no help from telepathy.

What we are told is that in Professor Rhine's experiments the percipient, while trying to guess the card that is dealt, happens, to a significant degree of frequency, to hit by mistake not that card but another card which has not been dealt, but is going to be dealt a few seconds later, which he is not trying to guess, and which is at the time not known to the dealer or any other human being. It is possibly made a little more plausible when we find that the guesses which do not hit their real mark are chiefly apt to hit the card next before or next after. This seems like normal shooting, a few shots hitting the bull, but more going just to the right or just to the left, but it would still involve some, to me, incredible hypotheses. I look hopefully towards the metaphysicians, such as Professors Broad and Price, to reveal some explanation which does not involve one of two incredible hypotheses: one that a man's naming of a card in an erroneous guess at another card should make a card of the sort named in another room make its way unobserved out of a pack and get itself dealt; the other, that the dealing of a card at a later time should cause a right guess to have been made some time earlier. To Dr. Soal it would make all the difference if at the later time the dealer should see the card; then it would at least not be the dead material card itself that caused the guess to have been made, it would be the thought which the dealer did not have at the time but was presently going to have which had influenced the mind of the guesser. Both views accept the conclusion which I find inacceptable, that an effect can precede its cause. I cannot feel much comforted by the explanation that the difference in time is only a matter of seconds and might be covered by that enduring moment which we commonly call 'now' or 'the present'. However, I know I may be wrong.

I feel on different grounds a similar incredulity about Dr.

PRESIDENTIAL ADDRESS

Rhine's startling cases of telekinesis. Considering the vast experience of the human race in tossing dice and coins and the extreme interest which millions of gamblers have taken for hundreds of years in the way they fall, if human thought or will could really compel a die or a coin to fall in the position the agent wishes, I think we should have heard much more about it by now. Here, again, I note with relief that Dr. Soal, too, is sceptical about telekinesis, and 'sees no sign of a genuine physical medium on the horizon'.

As to telepathy, however, I cannot maintain this healthy scepticism. There are three numbers of the *Proceedings S.P.R.* which would confound me if I did. They contain accounts and criticisms of my own experiences as a percipient; my Presidential Address in 1915, Mrs. Verrall's 'Report of a series of experiments in Guessing' (1916), and Mrs. Sidgwick's 'Report on Further Experiments in Thought Transference' in 1924. I have also several bundles of records of later sessions, though of late years, owing partly to the complete dispersal of my children and the rest of our old group, I have given up the experiments.

Let me say at once that my experiments belong to the pre-statistical stage of psychical research, when the experiments were treated almost as a parlour game. Still I do not see how there can have been any significant failure of control; nor did Mrs. Verrall or Mrs. Sidgwick. The conditions which suited me best were in many ways much the same as those which professional mediums have sometimes insisted upon. This is suspicious, yet fraud, I think, is out of the question; however slippery the behaviour of my subconscious, too many respectable people would have had to be its accomplices. I liked the general atmosphere to be friendly and familiar; any feeling of ill-temper or hostility was apt to spoil an experiment. Noises or interruption had a bad effect. One question that arose was the degree to which the telepathy made use of real sights, sounds, smells, memories, to reach its goal. The general conclusion was curious. It seemed that I, or my subconscious, showed some anxiety to explain away the telepathy by seizing upon some such excuse. It said it had guessed Savonarola making the women burn their precious possessions because it smelt a coal which had fallen out of the fire; that it had guessed Sir A. Zimmern riding on a beach in Greece because it said it had heard a horse on the road—when the rest of the company heard no horse. Memories, again, sometimes helped it, but more often hindered it in its search. At one time, indeed, I was inclined to attribute the whole thing to subconscious auditory hyperaesthesia. I got almost no successes if the subject was not spoken but only written down. Two or three successes and at least one error could be explained by my having heard or misheard a proper name, e.g. by confusing Judge Davies and the prophet David. But, apart from other difficulties in this hypothesis, there were some clear cases where I got a point or even a whole subject which had only been thought and not spoken.

Of course, the personal impression of the percipient himself is by no means conclusive evidence, but I do feel there is one almost universal quality in these guesses of mine which does suit telepathy and does not suit any other explanation. They always begin with a vague emotional quality or atmosphere: 'This is horrible, this is grotesque, this is full of anxiety'; or rarely, 'This is something delightful'; or sometimes, 'This is out of a book,' 'This is a Russian novel,' or the like. That seems like a direct impression of some human mind. Even in the failures this feeling of atmosphere often gets through. That is, it was not so much an act of cognition, or a piece of information that was transferred to me, but rather a feeling or an emotion; and it is notable that I never had any success in guessing mere cards or numbers, or any subject that was not in some way interesting or amusing.

Let us consider what we mean by telepathy. I believe most of us in this Society are inclined to agree with Bergson that it is probably a common unnoticed phenomenon in ordinary life, especially between intimates. We all know how often two friends get the same thought at the same moment. Tolstoy, the most acute of observers, speaks of 'the instinctive feeling with which one human being guesses another's thoughts, and which serves as the guiding thread of conversation'.[4]

The point will be clearer if I take some typical examples of my own experiences, both successful and unsuccessful. I choose them from the unpublished bundles of which I spoke, which are later than Mrs. Sidgwick's collection.

The method was always the same. I was sent out of the drawing-room either to the dining-room or to the end of the hall, the

[4] *Childhood and Youth*, p. 141. In another place he says of Nekhludoff that, 'when a chord was struck in his mind, a chord in mine vibrated'.

door or doors, of course, being shut. The others remained in the drawing-room: someone chose a subject, which was hastily written down, word for word. Then I was called in, and my words written down. I may add that, out of the first 505 cases, Mrs. Verrall estimated the percentage as: Success, 33 per cent; Partial Success, 27·9 per cent; Failure, 39 per cent. But it may be remarked that as evidence for the presence of some degree of telepathy most of the partial successes are quite as convincing as the complete successes: this would produce something like 60 per cent evidential and 40 per cent non-evidential.

First, two perfectly ordinary cases, where the emotional atmosphere is obvious and strong, and then is developed into something more definite.

October 26 1924 (?).
My wife gave a subject:
M.H.M. 'This is not a nice thing. What Nansen was describing the other day of the church yard at Buzuluk, where there lay the great pile of corpses, numbers of children who had fallen dead in the night.'
I was summoned, and said:
G.M. 'This is perfectly horrible. It's the Russian famine. It is the masses and masses of bodies carted up every night in the Church yard at . . .' (The scribe did not catch the name.)
M.H.M. 'Any particular bodies?'
G.M. 'Oh yes, children. I associate it with Nansen's lecture here.'

Here memory came in as a help. The subject was an incident that I remembered. In the next it was an obstacle: that is, a remembered incident thrust itself in and had to be rejected before I could get the real subject. I should explain that my mother had a story that when she was at a school in France, she had been made to wear a placard labelled *'impie'*.

November 24 1929.
MRS. DAVIES (agent). 'Jane Eyre at school standing on a stool, being called a liar by Mr. Brocklehurst. The school spread out below her and the Brocklehurst family "a mass of shot purple silk pelisses and orange feathers".'
G.M. '. . . (I think of) my mother being at her French school, being labelled *"impie"* . . . I reject that. But a sense of obloquy. Girl standing up on a form in a school, and the school there, and

people coming in, and she is being held up to obloquy in some way or other.—A thing in a book certainly. I think they are calling her a liar. I get an impression of the one girl standing up and a group of people or a family coming in and denouncing her. I think it's English.'
Question. 'Colour of the people's dresses?'
G.M. 'I can't get the colour of the people's dresses.'

I take another with a very marked but extremely different atmosphere.

January 22 1928.
STEPHEN MURRAY. 'George Rickey and me riding the motor-bike past the inhabitants of Moulsford Lunatic Asylum, and one cheery-looking man with gold spectacles on his forehead barking furiously at us, like a dog.'
G.M. 'A curiously confused and ridiculous scene. You and someone on a motor-bicycle, and a scene of great confusion; . . . perhaps the bicycle is broken down. But there is a confused rabble and, I know it sounds ridiculous, but someone on all-fours barking like a dog.' (Then after a little encouragement) 'Are they lunatics by any chance?'

Then two where the atmosphere is fainter and more subtle. The first came on a bad evening after two or three failures, and I was inclined to give up.

MY DAUGHTER ROSALIND. 'I think of dancing with the Head of the Dutch Foreign Office at a *café chantant* at the Hague.'
G.M. 'A faint impression of your journey abroad. I should say something official; sort of official soirée or dancing or something. Feel as if it was in Holland.'

The second occurred on 14 May 1927.

R.M. 'I think of walking in the Park at Belgrade and meeting the English governess.'
G.M. 'I'm getting a different feeling. It's somebody who is in rather a state of mind. I should think escaped from Russia. You are meeting her in some curious country. Wait a bit! It's not any-one at Robert College or Constantinople College. It's some queer country where you seem to be alone, and you are meeting some sort of Englishwoman who has been driven out of Russia, and hates the place where she is. . . . Oh yes. I do remember. It's when you went out to Constantinople by the express alone, and met the English governess in the Park.'

The history and 'state of mind' of the English governess was correct, but had not been mentioned. I had some faint memory of the incident, The 'queer country' was Serbia.

Next I will take two cases where I received a feeling or thought that had not been spoken, and was not in my memory at all.

November 17 1924.

R.M. 'A scene in a book by Aksakoff, where the children are being taken to their grandparents, and the little boy sees his mother kneeling beside the sofa where his father is lying, lamenting at having to leave them.'

G.M. 'I should say this was Russian. I think it's a book I haven't read. Somebody's remembrance of childhood or something. A family travelling, the children, father and mother. I should think they are going across the Volga. I don't think I can get it more accurately. The children are watching their parents or seeing something about their parents. . . . I should think Aksakoff. They are going to see their grandmother.'

Note. They did just afterwards have to cross the Volga, and Rosalind said she had been thinking of that, though she did not mention it.

Much more curious is the next, though at first sight it is a mere failure.

May 15 1927.

EDITH WEBSTER. 'I think of the Castalian spring at Delphi and how we drank the water there.'

G.M. I don't think I shall get it. But I've got a slight feeling of atmosphere, as if there were something terrible going to happen; as if it were the night before something . . . an atmosphere of suspense.'

Note. R.M. commented: 'I had been thinking of saying goodbye to someone who was going (to the war) to be killed, Hugo? Rupert? I got the feeling of "This is the end".'

R.M. had not spoken. She had evidently intended or expected to give the next subject, but E.W. was asked instead.

I add another failure which is, I think, equally significant.

November 24 1929.

MARGARET DAVIES. 'Medici chapel and tombs: sudden chill: absolute stillness. Marble figures who seem to have been there all night.'

G.M. 'I wonder if this is right . . . I've got a feeling of a scene in my *Nefrekepta*, where the man goes in, passage after passage, to the inner chamber where Nefrekepta is lying dead with the shadows of his wife and child sitting beside him . . . but I think it's Indian.'

(My poem was translated from an Egyptian story; I suppose I felt the subject was not Egyptian.) Sometimes the subject was a bit of poetry: I was then apt to answer at once without any groping or hesitation.

January 22 1928.

MARGARET COLE. 'The man in Browning who is dying and sees the row of bottles at the bed, and it reminds him of where he met his girl when he was young.'

G.M. (Instantly on entrance.)
'How sad and mad and bad it was,
But oh, how it was sweet.'

JOHN ALLEN. 'I think of the priest walking by the shore of the sea after he had been to Agamemnon and been refused.'

G.M. Βῆ δ' ἀκέων παρὰ θῖνα πολυφλοίσβοιο θαλάσσης
[*Iliad*, 1. 34.]

Now, granted that this curious sensitivity which we call telepathy exists, how shall we best analyse or describe it? In the first place, as far as my own experience goes, it does not quite feel like cognition or detection; it is more like the original sense of the word 'sympathy', συμπαθεία, the sharing of a feeling, or 'co-sensitivity'. I seem to be passive, and feel in a faint shadowy way the feeling or state of mind of someone else. Tolstoy's metaphor of the chord which vibrates when another chord is struck seems to express it.

If we follow the general lines suggested by Bergson we may suppose an original store, so to speak, of vague undifferentiated sensitivity belonging to all gregarious creatures, which is then 'canalized' into particular clearer and more efficient forms as the creature develops definite sense organs. These senses again become keener and more effective as they are needed and used, but fade away if they are not used, while some remnant of the original weak uncanalized sensitivity is still there to be drawn upon.

The points on which I speak with some conviction are, first, that telepathic communication does take place, and secondly that as far as my own experience goes it seems to me to be a communication of feeling rather than of cognition, though the cognition may follow as the feeling is interpreted. To these I add the conjecture that our ordinary everyday telepathy may be a faded and greatly intellectualized form of a sensitivity which exists much more simply and widely among many birds and gregarious animals and primitive races of men.

But there may be more than that in it. The differences between the human and non-human are very great. Our whole range of sensitivity has been so widely increased by our possession of such tools as hands and language. We cannot see like a hawk or track like a dog or hear like a hunted deer; but we can see a Rembrandt picture and feel the thrill of a Beethoven sonata or a great poem. And surely it is noteworthy that just here our sensitivity passes beyond the realm of mere observation into that of feeling; beyond the facts that you observe there is the sense of other things, not fully known, which have value and importance. I have already noticed that our faculty of telepathy, such as it is, seems to operate best in just those spheres where our normal instrument, language, either fails or works with difficulty. It is certain, I suppose, that there still are more things in heaven and earth than are at present mastered by science. And Bergson has reminded us that millions of men have lived for thousands of years in a world vibrating with electricity, without ever suspecting that there was such a thing. Are we not probably now in the presence and under the influence of unknown forces, forces concerned with deeper or more remote values or beauties or loyalties, which are beyond the range of our exact knowledge and power of definition, but by no means beyond the reach of an undefined but strong and even passionate feeling: 'Thas is what I value,' 'This is what I love,' 'This is what I must obey'; or negatively, 'This is what I reject.' I suspect that what we call genius is a special sensitiveness in this region of art, poetry, thought, and the like: a sensitiveness which according to many critics is apt to be deadened and disregarded by our all-absorbing material civilization, and if so, is disregarded at our peril. It is in that region that our great tool, language, fails us and we have most highly developed our ancient pre-linguistic or supra-linguistic sympathy. If this is so, it may well be that William James was right in his forecast that the work of this Society may ultimately render great service to the religious gropings of the human mind.

III

PSYCHICAL RESEARCH AND HUMAN PERSONALITY*

H. H. Price

My aim in this paper is to show the relevance of Psychical Research to certain questions which interest all reflective persons, and have always interested them—in one form or another—in all ages and countries.

Many people, even highly educated and highly intelligent people, still speak and write as if Psychical Research did not exist. Many people think that it is the same thing as Spiritualism, whereas Spiritualism is really just one hypothesis among others to account for certain of the phenomena which Psychical Researchers investigate. And some people who make neither of these mistakes still seem to think of the Psychical Researcher as a collector of rather queer facts which have no particular importance. Certainly the facts *are* queer (perhaps there is nothing in our whole experience which is queerer than Precognition). But they are not on that account unimportant. On the contrary, if they are genuine— as I am sure many of them are—they make a fundamental difference to our whole outlook: to our conception of human personality and of its place in the universe. That, at any rate, is what I hope to show.

In our Western society, the traditional conception of human personality was a dualistic one. The classical expression of it is to be found in the philosophy of Descartes. It was thought that the human being is a compound of two wholly different but interacting substances, mind or soul on the one hand, body on the

* Reprinted from the *Hibbert Journal*, Vol. XLVII, No. 2, January 1949.

other. With the advance of scientific knowledge, this theory has come to seem less and less credible. It is not that any empirical facts have been discovered which conclusively refute it. But can one deny that the appearances are against it? The biological sciences find no evidence in support of psycho-physical interaction. On the contrary, they suggest that mental processes of every kind are unilaterally dependent upon physico-chemical processes in the brain. No doubt it is true that each of us 'has a mind', if you mean by this that mental events do occur in connection with every living human body. But the notion that there is a mental substance or soul-thing, the *res cogitans* of Descartes, to which these mental events 'belong', has come to seem absurd. What they 'belong to', it would be said, is just the living human organism, as physiological processes like digestion also belong to it. This, or something like this, is the Materialistic conception of human personality, technically called Epiphenomenalism. It has come to be accepted by the vast majority of educated Western people; or rather, it is not so much accepted by them (for this would suggest a conscious choice between one theory and another, and that stage has long been passed), but has come to be taken for granted as something obvious and beyond dispute. What interests them now are not the arguments in favour of this conception of human nature, but its consequences; the secularistic or 'this-wordly' outlook which it entails, and the transvaluation, or devaluation, of traditional values which follow from this.

The Materialistic theory of human nature (or the Naturalistic theory if that term be preferred) is certainly a most impressive one. It draws its strength from a large mass of well-established empirical facts. Yet it has one weakness. It confines itself to the facts of 'normal' experience. But these are not the only relevant data there are. We must take the 'supernormal' phenomena into account as well. The Materialist appeals to the empirical facts, and rightly so. 'Very well,' we must say to him, 'thou hast appealed to Caesar: to Caesar thou shalt go.' There are empirical facts, and highly relevant ones too, which he has failed to consider at all.

It is obvious at once that if there were conclusive empirical evidence for the persistence of human minds after bodily death—evidence as conclusive as that which convinces us of the continued existence of human beings who have emigrated to Australia—then the Materialistic conception of human personality would be directly and finally refuted. If the human mind continues to exist after the disintegration of the brain, it cannot be true that all mental events are unilaterally dependent upon physical events in the brain, as the Epiphenomenalist says they are. Now, of course, there is a good deal of empirical evidence which supports the hypothesis of human survival; most of it is derived from mediumistic communications, which are among the most important of all supernormal phenomena. I do not, however, think that this evidence is as yet absolutely conclusive, though it does not follow that it never will be. I know that many investigators, whose opinion I respect, will disagree with me; they hold that the evidence for survival is conclusive already. I would suggest, however, with all deference, that they have not taken sufficient account of the very queer facts established by Abnormal Psychology concerning dissociated personality and the possibility of secondary personalities. I think that they are taking an oversimplified view of the human mind (the mind of the medium in this case); in fact, the view of Descartes, who regarded the human mind as a simple thinking substance. I shall come back to this point later; I believe it is highly relevant to all the problems of Psychical Research, and certainly it is relevant to this one. We must face the possibility that what appears to be an extraneous and discarnate personality, manifesting itself by means of the vocal or other organs of the medium, may in fact be a secondary personality of the medium herself. Such secondary personalities can show a surprising degree of autonomy and internal cohesion, as the phenomena of Psychopathology make clear; they can also show a surprising degree of intelligence and purpose. There is no reason why they should not possess supernormal cognitive powers, such as telepathy and clairvoyance. Indeed, there is some reason to think that such powers are more likely to operate freely when the mind is in a dissociated state and the normal personality is in abeyance. In view of these troublesome complications—and it really will not do to ignore them—I would only venture to go as far as this: I think that there are some mediumistic communications which are very difficult indeed to explain on any other hypothesis except the hypothesis of survival, but I do not think that there are any which absolutely prove it. The man who denies survival is certainly on dangerous ground. But I think that he still has a leg, or half a leg, to stand on. And if I am right, this 'short way with

Materialism'—his direct and knockdown method of disproving it—is not open to us, at any rate at present.

But though the short way is not open to us, there may be a longer way which is. Mediumistic communications are not the only sort of supernormal phenomena, important as they may be. I think that if we consider the implications of Telepathy, the most elementary and the best established phenomenon in the whole field of Psychical Research, we shall see that they are incompatible with the Materialistic conception of human personality. In the remainder of my paper I shall try to show that this is so. If my argument is correct, it will, of course, follow that the antecedent *im*probability of the survival hypothesis (an improbability derived from the facts of normal experience, especially the findings of the biological sciences) is greatly diminished, once the existence of telepathy is admitted, as I think it certainly must be.

In telepathy one mind affects another without any discoverable physical intermediary, and regardless of the spatial distance between their respective bodies. The Materialist, once we can get him to admit the facts, will no doubt try to explain them by physical radiations of some kind. Indeed, that is what he *must* do if his conception of human personality is to stand. But no explanation of that kind seems to be feasible. Such physical radiations, if they exist, ought to be detectable by physical instruments. It ought to be possible to intercept them *en route*; and their intensity should vary in some way with the spatial distance between the body of the agent and the body of the percipient. But none of these consequences, which ought to follow if the Radiation Theory is true, is in fact verified. If the supposed physical radiations do not have any of the empirically verifiable properties which physical radiations have, what is the point of calling them radiations at all, and what is the point of calling them physical either? They cannot be physical in the *ordinary* sense of the word 'physical', the one which the Materialist is using when he says that all mental events have physical causes. For in the ordinary sense of the word 'physical' nothing is a physical event or entity unless it is perceptible by means of the sense-organs; either directly, or indirectly by means of instruments which can themselves be directly perceived. This *is* true of events in the nervous system, the events of which the Materialist was speaking when he formulated his Epiphenomenalistic theory of human personality; but it is not true of the hypothetical physical radiations which are alleged to be the cause of telepathy. They have just been postulated *ad hoc*, because no physical explanation of telepathy is possible in the ordinary sense of the word 'physical'. And all that the postulate really amounts to is this: 'something or other must be happening in space when telepathy occurs, but we have not the ghost of an idea what it is'.

For my part, I see no empirical reason for making any such postulate. But if one does make it (as I know that many people are inclined to) I do not think it really supports the Materialist's case at all. What it would support is something very different—what I will call the *Occultistic* conception of human personality. According to the Occultists, every human being has several 'higher' bodies, in addition to the ordinary physical body which our senses make us aware of; and each of these higher bodies responds in its appropriate way to a 'higher' sort of physical environment, which likewise differs from the ordinary physical environment revealed by our senses. It seems to me that the postulated spatial processes, 'thought waves' and the like, together with the organs for emitting and receiving them, will have to exist in one of these 'higher' physical worlds if they exist at all, since we cannot find them in this one.

Now, of course, we know very little about the universe, and it may well be a much queerer place than most of us think. It is theoretically conceivable that there might be such higher bodies, and higher worlds in which they function. But I see no reason to believe it, unless or until their existence can be empirically verified (presumably by some sort of clairvoyance). I suppose that there could be a kind of Occultistic Epiphenomenalism—perhaps there is—a theory which holds that all mental events are unilaterally dependent upon events occurring in one or other of these superphysical bodies. But such a theory, what one might call 'Materialism with a higher kind of matter', would still be quite incompatible with *ordinary* Materialism, which is the only kind that concerns me in this paper. For according to the Occultists these higher bodies can and do continue to exist when the ordinary physical body is destroyed.

However this may be, I think that in our present state of ignorance we should stick to the ascertained facts. And these suggest a purely 'mental' theory of telepathy. They suggest that in telepathy one mind affects another *directly*, and that nothing whatever travels through space between their respective bodies or brains. Such a type of causation seems to be quite inconsistent with the Materialistic scheme. As we have seen, the Materialist will not allow that there is any such thing as purely psychological causation even *within* the individual mind. (The causal linkage, in his view, is a linkage between the 'underlying' physical events in the brain, and not between one mental event and another.) If an experience in my mind, for example a dream, is directly caused by an event in another mind, without any intermediate causal linkage between our respective brains, then it cannot be true that all mental events are wholly caused by brain events, as the Materialist says they are. He may indeed point out that all the telepathic experiences we know of have occurred in *embodied* minds; and this, he may say, suggests that the presence of a living and normally functioning brain is a necessary condition of their occurrence. But even if we allow that he is right in this, it still does not give him what he needs. The presence of a living and normally functioning brain is a general precondition of all experiences whatever, at least in the embodied mind; but it does not explain any one experience in particular. In the case of telepathy there is not that *detailed* correlation between mental events and brain-events which the Materialist theory requires—the kind of correlation which we do find between brain-events and visual sensations, for example.

We must conclude, I think, that there is no room for telepathy in a Materialistic universe. Telepathy is something which ought not to happen at all, if the Materialistic theory were true. But it does happen. So there must be something seriously wrong with the Materialistic theory, however numerous and imposing the *normal* facts which support it may be.

It is to be noticed, however, that although telepathy does not fit in with the Materialistic conception of human personality at all, it does not altogether fit in with the traditional religious conception either, at any rate if we confine our attention to the religious tradition of Western Europe. For the traditional religious conception of human nature is not only dualistic, regarding mind and body as two different and separable entities. It is also, if I may say so, an 'isolationist' conception with regard to the individual mind. It holds that each individual mind is a separate and complete substance, whose only direct causal relations with the rest of the universe (apart from God) are relations with its own brain.

The individual mind, it is supposed, can affect and be affected by other finite minds only in a very indirect and circuitous manner, by a long intervening chain of physical causes. The existence of telepathy shows that this 'isolationism' is false, even with regard to embodied minds; and *a fortiori* false with regard to disembodied ones, if there are any. It is not true that the only part of the universe with which a given mind has direct causal relations is its own body. It also has them with other minds. (If Telekinesis is ever established as a genuine fact, we shall have to say that it has them with other parts of the physical world as well.)

Moreover, and equally important, when the traditional religious theory of human personality maintains that the individual mind is a psychical substance, we have to point out that the notion of substance does not seem to fit the empirical facts. Here, as elsewhere, Psychical Research joins hands with Abnormal Psychology. The phenomena of dissociated and alternating personality seem to show that the individual human mind does not have the internal unity which the notion of substance requires; and they certainly show that it is not the *simple* substance which the old Dualistic philosophers thought it to be. On the other hand, the phenomena of telepathy show that one mind is not separated from another by any sharp and clear-cut boundary. Imagine two minds which were in a state of complete and continuous telepathic *rapport*, so that every experience of either directly affected the experiences of the other. Would there any longer be any sense in calling them two minds and not one? If the causal connection between two sets of mental states were as close as this, we should have to say that there was one mind in two bodies; just as, if there is a sufficient degree of *dis*connection between two groups of mental states both of which are associated with the same body, we have to say that the mind animating that body has been split into two separate personalities.

It comes to this: both *ad intra* and *ad extra* (if I may so put it) the unitariness of the human mind seems to be a matter of degree, and not a matter of all or none. In relation to other minds, it is without clear boundaries; and its internal coherence is greater or less in different circumstances, and never perhaps complete (some degree of dissociation occurs in all of us whenever we dream). In view of these facts, it is quite inappropriate, indeed positively misleading, to say that the human mind is a substance.

I think that in the philosophy of mind, as in the philosophy of matter also, the time has come when we must throw overboard the heritage of Descartes. The concepts devised by that illustrious man have formed as it were the intellectual working capital of educated Europeans for three centuries, both in their thoughts about the physical world and in their thoughts about the human mind. But now, in both spheres alike, they are becoming a nuisance. It seems to me that one of the greatest obstacles to the understanding of supernormal phenomena is precisely the Cartesian notion of the human mind as a psychical substance or *res cogitans*. The phenomena do not make sense in a Cartesian universe, no more than they make sense in a Materialistic one. They make no sense, because we are choosing the wrong unit, so to speak; we are trying to map out the psychological world into so many distinct and separate individual minds, and assuming that every mental event must be attributed to one or other of them. But this way of mapping out the psychological world does not fit the facts. Is the 'control' of a medium an individual mind or not? If a haunting apparition displays a certain degree of intelligence and purpose, but not very much, are we to say that it is a manifestation of an individual mind or not? If we are to talk intelligently about such queer entities, if we are even to ask intelligible questions about them, I believe that we must change the unit, as it were. We must take as our fundamental unit something far less complicated than a complete mind, something like an individual idea, and build up the various grades of psychical entity out of them: from not-very-purposive ghosts and Freudian complexes at the one end, to the complete and healthily integrated human mind at the other, with mediumistic 'controls' somewhere in the middle. All these different sorts of mental entity, we must say, and any others there may be, are idea-systems of different degrees of complexity and different degrees of autonomy and internal coherence.[1]

I do not want to suggest, however, that Descartes was entirely wrong when he introduced the notion of a *res cogitans* or conscious substance. Human personality is a very complex thing, and there

[1] This is very like the late Mr. Whately Carington's Psychon Theory. But his fundamental unit is a mental image rather than an idea, and he would have objected very strongly to the qualifications introduced in the next paragraph. Cf. his book *Telepathy*, pp. 96 ff.

may be some factor in it which does deserve this honourable title. But if anyone thinks that there is, I believe he would be well advised to go behind Descartes to an older tradition, which divides human nature into *three* parts, body, mind (or soul, ψυχή) and spirit, instead of Descartes' two, body and mind. And one would then say that it is the spirit or pure ego which is a substance, though the mind or soul is not. This tripartite division of human nature appears under various names in Neoplatonism, in some of the religious philosophies of the Far East, and in some Christian thinkers also. There are philosophical arguments in favour of it which have some weight; moreover, certain forms of mystical experience seem to support it. If we do accept this threefold division, we shall have to say that it is mind or soul, and not spirit, which is the subject-matter of the psychological sciences—Normal Psychology, Abnormal Psychology, and the Supernormal Psychology which we now call Psychical Research. The thing which does *not* deserve to be called a substance, the thing which has no clear boundaries and whose internal unity is a matter of degree, will be mind or soul, according to this terminology. In the study of telepathic phenomena, we are investigating the fuzziness of its boundaries, the way in which one mind or soul overlaps with others; in Abnormal Psychology, we are studying its internal coherence and the conditions which cause that coherence to break down.

Such considerations, however, take us far away from the Materialistic conception of human personality, and we must now return to it. For the phenomena of telepathy are relevant to the truth or falsity of the Materialistic conception in another way, which I have not yet mentioned. It seems pretty clear that the telepathic 'contact' of mind with mind (or the telepathic 'overlap' of one mind with another) is something which occurs in the first place at the unconscious level; in the terminology of F. W. H. Myers, it occurs in the subliminal region of our minds, beneath the threshold of consciousness. It looks as if telepathically received impressions had some difficulty in crossing the threshold and manifesting themselves in consciousness. There seems to be some barrier or repressive mechanism which tends to shut them out from consciousness, a barrier which is rather difficult to pass, and they make use of all sorts of devices for overcoming it. Sometimes they make use of the muscular mechanism of the body, and emerge in the form of automatic speech or writing.

Sometimes they emerge in the form of dreams, sometimes as visual or auditory hallucinations. And often they can only emerge in a distorted and symbolic form (as other unconscious mental contents do). It is a plausible guess that many of our everyday thoughts and emotions are telepathic or partly telepathic in origin, but are not recognized to be so, because they are so much distorted and mixed with other mental contents in the process of crossing the threshold of consciousness. It is also a plausible guess that we receive many telepathic impressions which never reach consciousness at all; or if they do, reach it only in the form of a vague 'tone' or 'colouring' pervading our consciousness as a whole (a kind of mass-effect as it were, in which individual items cannot be distinguished).

Now it might be argued that in the Materialistic conception of human nature the only mental events which are recognized at all are those which occur within consciousness, and that the subliminal region of our personality is ignored altogether. If so, telepathic phenomena would be inconsistent with the Materialistic conception in another way, since we cannot make any sense of them unless we take subliminal mental events into account. Of course, if this argument were valid, the phenomena of Abnormal Psychology would be equally incompatible with the Materialistic theory; for we cannot make any sense of them either, so long as we suppose that the only mental events there are those which occur within consciousness.

It must be admitted, I think, that some Materialists *have* ignored the subliminal or unconscious strata of human personality, and *have* spoken as if the only mental events which occur at all were those which occur within consciousness. If so, they were mistaken and their conception of human personality was far too simple to be true.[2] Nevertheless the Materialistic theory can perfectly well admit the existence of unconscious or subliminal mental events, even though its advocates have not always seen that it can. The brain is a very complicated structure indeed. All kinds of other physical processes go on in it besides the ones which (in the Materialist view) produce conscious mental states; and some of these other physical processes might have unconscious mental

[2] I am assuming here that a purely *physiological* theory of the Unconscious is inadequate. It is certainly inadequate if our physiology is mechanistic, as Materialistic physiology is.

states as their by-products. Moreover, we must remember that physical processes may have different levels of complexity. It might be that processes at the sub-atomic level produce unconscious mental events, while processes at the atomic or molecular level produce conscious ones. Thus there is plenty of room in the brain for physical correlates of unconscious mental events, if anyone wants them. We must remember also that the intrinsic nature of subliminal events is quite unknown to us. We only call them mental at all because of the nature of their effects. (We say, for example, that Smith's behaviour is caused by an unconscious hostility towards Jones, because he acts or speaks or dreams *as if* he wished to injure Jones, but is not in fact conscious of any such wish.)

Thus there is no logical absurdity in a Materialistic theory of the subliminal mind. Some students of Abnormal Psychology do in fact profess to be Materialists; and we need not suppose that this is just a face-saving inconsistency, designed to placate scientific orthodoxy. The objection to this extension of the Materialistic theory is not that it is in any way inconsistent with the Materialist's premises. It is rather that an Epiphenomenalist conception of the subliminal mind is completely *unfruitful*. It is altogether too vague and general to throw any light on the empirical facts. It does not enable us to explain the phenomena of either Abnormal or Supernormal Psychology in detail, or to make verifiable predictions, in the way that the Freudian theory of dreams, for example, does. On the other hand, we find that we *can* to some degree explain and predict the phenomena if we conceive of unconscious events in a purely psychological manner, without any attempt to correlate them in detail with 'underlying' physiological occurrences. In fact, a Materialistic theory of unconscious mental events has the defect which some metaphysical theories have; by explaining everything in general it explains nothing in particular. Indeed, it *is* a metaphysical theory in the dyslogistic sense of the word 'metaphysical', whereas the Materialistic theory of conscious mental events—right or wrong—is an empirical one.

I have said a great deal about telepathy, too much perhaps. But, of course, there are other supernormal phenomena which are equally difficult to reconcile with the Materialistic conception of human personality. Precognition is one. Precognition seems to require a mode of causation in which the effect occurs *earlier* than the cause, and there is clearly no room for such a process in a Materialistic universe. But I should not like to lay too much stress on this, because it is very hard to find room for such a paradoxical form of causation in a non-Materialistic universe either. It is difficult to think of any hypothesis at all which will make precognition intelligible. The most hopeful perhaps is the suggestion that there is some other kind of time (or some other dimension of time) besides the one we are familiar with in our normal experience. If there is such a non-physical time or time dimension, and if some human minds have access to it, this again is clearly something which does not fit in with the Materialistic philosophy in any way.

What is to be said about Clairvoyance, the veridical cognition of contemporary physical objects or events without the use of any sense-organ or any process of rational inference? Some recent investigators have maintained that the evidence for clairvoyance is quite inconclusive. They hold that the empirical facts can all be accounted for by telepathy or a combination of telepathy and precognition (though Dr. Rhine does not agree). I am not altogether persuaded that they are right. Certainly it is difficult to isolate clairvoyance from telepathy under experimental conditions, and experimental tests for pure clairvoyance have several times yielded negative results. But there are still the spontaneous cases of clairvoyance, and I doubt whether all of these can be wholly ex-explained by telepathy and/or recognition, though some of them probably can be. If I am right, and if there is after all such a thing as pure clairvoyance, however difficult it may be to isolate, then here again is a phenomenon, which is irreconcilable with the Materialistic conception of human personality. According to the Materialist theory, all our knowledge of the external world comes to us indirectly, through a long chain of physical causation, by means of physical stimuli affecting our sense-organs, which in turn cause physical changes in our nervous systems and brains. But clairvoyance involves some sort of *direct* relation between the mind and the external world. No doubt clairvoyance, like telepathy, is something which occurs in the first place at the subliminal level, and manifests itself in consciousness indirectly, often in a symbolic form. All the same, the subliminal mind itself must be in a direct relation with physical objects or events outside the body which has nothing to do with the physical sense-organs.

It is, however, sufficient for my argument if the reality of telepathy is admitted. Telepathy, as I have said, is something which ought not to occur at all if the Materialistic conception of human personality is correct. But it undoubtedly does occur. The Materialistic conception of human personality must therefore be mistaken, even though all the facts of *normal* experience seem to favour it, and even though the philosophical arguments against it are inconclusive.

In conclusion, I would again like to address myself to the people whom I mentioned at the beginning (I believe they are fairly numerous); to those who agree that Psychical Research has succeeded in establishing various queer facts about the human mind, but think that these facts are mere curiosities and oddities, of no particular importance. Certainly card-guessing does appear at first sight to be a rather trivial occupation. And if a few dreams turn out to be telepathic or precognitive, if people do occasionally have veridical telepathic vision, why should anyone make such a fuss about it? On the contrary, these queer facts are not at all trivial, and it is right to make the greatest possible fuss about them. Their very queerness is just what makes them so significant. We call them 'queer' just because they will not fit in with orthodox scientific ideas about the universe and man's place in it. If they show, as I think they do, that the Materialistic conception of human personality is untenable, and if they throw quite new light on the age-old conflict between the scientific and the religious outlooks, we shall have to conclude that Psychical Research is one of the most important branches of investigation which the human mind has ever undertaken.

IV

NOTES ON CHANGING MENTAL CLIMATES AND RESEARCH INTO ESP

Rosalind Heywood

The course of research into ESP can only be understood in the light of its historical background and of the mental and emotional climate in which it has been undertaken.

In the East, on the whole, ESP has always been accepted as a genuine phenomenon, and before the Age of Science this was also true of the West. In Biblical times it occasioned no surprise that the Prophet Elisha was able to tap the war plans of the King of Syria by telepathy, or that the Delphic Oracle could describe what King Croesus of Lydia was doing, hundreds of miles away, at the time she spoke. But it was to test her skill that Croesus challenged the Oracle to divine his occupation: he would never have asked *how* she did it, for divination—ESP—was supernatural, of the Gods, and not a matter into which mortals cared to pry.

This connection of ESP with religions which the Christian Church supplanted was an obvious reason for her ruling that, except in very special cases, it derived, not from the Deity but from the Devil, and was therefore to be avoided under ban of her severe displeasure. Nevertheless, apparent spontaneous instances of it continued to occur, and since it was an excellent source of power over the ignorant, and hence of profit, it also continued, *sub rosa*, to be faked. Nor did serious interest in it ever die out, although it was driven underground by the Church, and from classical times until the nineteenth century we hear of groups of intellectuals—Neo-Platonists, Alchemists, Rosicrucians, Cambridge Platonists, and others—getting together in secret to indulge in occult practices. The populace, too, believed obstinately in witchcraft, and some witches may have been genuinely gifted with ESP, although today many of their phenomena would be recognized as hysterical symptoms, or, possibly, as the effects of eating hallucinogenic plants or mushrooms. But the intellectuals were not scientific investigators: what they sought was illumination or power; and the poor were after no more than material help.

With the coming of the Age of Science, ESP fell yet farther from its high estate: in the eyes of the educated it ceased to be even of the Devil; it ceased, officially, to exist at all. The universe as seen by Newton's successors, ordered, predictable, built up of tiny billiard-ball atoms which whirled around in the aether according to immutable laws of cause and effect—this held no place for information arriving by no apparent means from nowhere. 'Philosophy is often much embarrassed,' said Kant, on hearing that his brilliant contemporary Swedenborg had mysteriously become aware of a distant fire at the actual time it was burning, 'when she encounters certain facts which she dare not doubt, yet will not believe for fear of ridicule.'

But, as time went on, practically all scientifically educated persons found that fear of ridicule, plus their own very reasonable recoil from the seemingly irrational, was more powerful than alleged facts which did not fit into the scheme of things; so, humanly enough, like the man who refused to look through Galileo's telescope for fear that what he saw would not suit his views, they safeguarded themselves by ignoring the evidence for ESP. The nineteenth-century scientist Helmholtz expressed this attitude with dramatic force. 'Neither the testimony of all the Fellows of the Royal Society nor the evidence of my own senses,' he declared, 'would lead me to believe in the transmission of thought from one person to another independently of the recognized channels of sense.' The great Lord Kelvin, too, pronounced that such wretched superstitions as ESP, hypnotism, and the like 'were only fit for the rubbish heap'.

However, in spite of being officially non-existent, ESP still appeared to crop up, until at last, in the fourth quarter of the nineteenth century, a small group of adventurous-minded scholars and scientists, mostly from Cambridge, were driven to ask the heretical question: Can there be quite so much smoke with no fire behind it? Is it possible that supernatural phenomena do occur and are not supernatural after all, but normal and subject to natural law?

This question was the seed from which modern psychical research (or parapsychology) has grown. Its first-fruit was the foundation of the British Society for Psychical Research, with Professor Henry Sidgwick as its first President, and Lord Rayleigh, Sir William Barrett, Sir William Crookes, Professor J. J. Thomson, the first two Lords Balfour, Frederic Myers, and Edmund Gurney among its early members. Their aim was to turn the searchlight of cold-blooded, co-ordinated scientific inquiry on to the hitherto uncanny, supernatural, non-existent ESP and related, equally unthinkable, phenomena. Soon afterwards a sister Society was formed in America, with William James as a member, and in Europe, over the years, a few first-class thinkers also began to look into the subject, among them Charles Richet, Flammarion, Bergson, Pierre Janet, and Hans Driesch.

It so happened that the S.P.R. was founded at a time when classical Newtonian physics was at its zenith. It is recorded that in 1887 an eminent chemist, Marcellin Berthelot, said that 'from now on there is no mystery left in the universe', and that a few years later the physicist Professor Gabriel Lippmann told a brilliant pupil that he had better take up some other line of research, as the subject of physics was exhausted, over, and done with. Sidgwick and his colleagues were well aware that to found a Society for the study of ESP in such a mental climate was to ask for trouble.

> My highest ambition in psychical research [he said in his first Presidential address] is to produce evidence which will drive my opponents to doubt my veracity. I think there is a very small minority of persons who will *not* doubt it, and that, if I can convince them, I have done all that I can do. As regards the majority, even of my own acquaintance, I should claim no more than an admission that they were considerably surprised to find *me* in the trick.

Sidgwick was right: trouble came to the psychical researchers in full measure. 'We were told somewhat roughly,' he wrote in later years, 'that, being just like the other fools who collected old women's stories and solemnly recorded the tricks of imposters, we only made ourselves the more ridiculous by assuming the aims of a scientific society and varnishing this wretched nonsense with technical jargon.'

Such criticism seems to spring from emotion as well as from cool intellectual dissent from foolish ideas, and reasons for that are easy to understand: disgust at ESP's age-old entanglement with the fraud always invited by those who 'hunger and thirst to be bamboozled', atavistic fear of the supernatural, which can exist undetected beneath the most rational mind, and as well, perhaps, unconscious fear of a phenomenon which might imply invasion of the personality by a flood of impressions which could not be warded off. There was also, doubtless, a very human suspicion of out-of-the-ordinary ideas in general, especially of those which could undermine the basis of one's own work. 'If that were true,' said a well-known mathematician many years later, of ESP, 'it would mean that I would have to scrap everything and start again from the beginning.' Even today there are people who seem driven to spend much of their time grubbing for mud to sling at both past and present psychical researchers, apparently because, for some obscure motive, they *want* their integrity to be suspect. The springs of over-emotional resistance to ESP in the educated are as interesting as the widespread over-credulity towards it among simpler persons.

Largely owing to the leadership of Sidgwick, 'whose reputation for sanity, truthfulness and fairness', says Professor C. D. Broad, 'was well known to everyone who mattered in England', the hostile mental climate did not actually destroy the S.P.R., but it did, perhaps, canalize its activities along a somewhat narrow, though obviously vital, channel: to inquire by the methods of science whether or not ESP existed at all. These methods, when studying a new phenomenon, are well known: observe it as often and as objectively as you can; put forward a tentative hypothesis for it, and learn to reproduce it in controlled conditions as and when required. Such a procedure is comparatively easy in the physical sciences, less so in biology, less so again in psychology, and immensely difficult when the quarry is as elusive as ESP. This elusiveness was—and is—the root of the trouble. The early researchers collected hundreds of apparent examples of it which were well corroborated; they put forward tentative hypotheses for it; they did pioneer experiments in which it appeared sporadically to crop up. But they did *not* succeed in producing it in controlled conditions to order.

Apart from experimental work, they had two main sources of material: (i) upwellings from the subconscious of certain persons, including spiritualist mediums, by means of automatic speech or writing, which sometimes contained information which could not have been known to those persons; and (ii) spontaneous ESP-type experiences reported by the public. These spontaneous experiences follow roughly one of two patterns: either B gets an impression of an event or state of mind occurring to a distant A, or B feels aware of A's presence or hears his voice when A is physically elsewhere. To accept such experiences as evidential the researchers required two things: confirmation from a third person that B had reported his impression *before* he could have heard of the related event by normal means, and, for comparison, reliable information about the event itself.

Critics argued that this type of evidence, however well attested, could *never* be good enough, because, fraud and exhibitionism apart, it always contained a subjective element, and such an element was particularly suspect in relation to ESP as this so often involved emotion. Moreover, facts were not always noted and confirmed at the time, and memory, even when emotion is not involved, is notoriously unreliable. And in theory, even the best attested correspondences between an ESP-type impression and a related event could always have come about by chance. 'Take every alleged case of ESP *separately*,' said the critics, 'and it can easily be explained away as due to some such normal cause. And ten thousand leaky buckets hold no more water than one.'

Sidgwick and his colleagues did indeed learn by experience that many apparent cases of ESP could be ascribed to bad observation, wishful thinking, unconscious inference, false deduction, faulty memory, or chance coincidence, but years of careful investigation convinced them that there remained a large and solid core of evidence for it which could not, in honesty, be got rid of in this way. Yet the scientific world as a whole still rejected it: *nothing* is good enough but ESP to order in the presence of hostile critics, continued to be the attitude, and that this could not be achieved confirmed the view that it was an absurdity which did not exist. Indeed it became a symbol of the absurd. Before the Heaviside layer was discovered critics jeered that, since radio waves go straight and the world is round, Marconi's idea of radio communication between Cornwall and Newfoundland was as silly as that of telepathy. And at the turn of the century Professor Joseph Jastrow labelled the idea of telepathy an 'egregious logical sin'. No one envisaged that, by its very nature, it might be inhibited by the presence of hostile critics, or that, before it could function, unknown and complex physiological and psychological conditions, and even inter-personal relations, might all have to be in a favourable state at the same time. Take, as a possible analogy, an electric lamp. For it to function, bulb, switch, plugs, fuse, wiring must all be intact and connected up; a 'short' anywhere along the line between lamp and generating station, and there will be no light. Similar 'shorts' may prevent the emergence of ESP; if they are intermittent it may flash on sporadically and unpredictably.

Over and above the reasons listed for hostility towards psychical research, there was a very powerful one provided by the researchers themselves. That was their own non-committal attitude towards the nature of man, which could scarcely be popular with either side at a time when the battle between traditional religion and scientific materialism was at its height. This is how Sidgwick expressed their attitude:

> When we took up seriously the obscure and perplexing investigation which we call psychical research, we were mainly moved to do so by the profound and painful division and conflict as regards the nature and destiny of the human soul which we found in the thought of our age. On the one hand, under the influence of Christian teaching, still dominant over the minds of the majority of educated persons, and powerfully influencing many even of those who have discarded its dogmatic system, the soul is conceived as independent of the bodily system and destined to survive it. On the other hand, the preponderant tendency of modern physiology has been more and more to exclude this conception, and to treat the life and processes of any individual mind as inseparably connected with the life and processes of the short-lived body that it animates. . . . Now our own position was this: we believed unreservedly in the methods of modern science, and were prepared submissively to accept her reasoned conclusions, when sustained by the agreement of experts; but we were not prepared to submit with equal docility to the mere prejudices of scientific men. And it appeared to us that there was an important body of evidence—tending *prima facie* to establish the independence of the soul or

spirit—which modern science had simply left on one side with ignorant contempt; and that in so leaving it she had been untrue to her professed method and had arrived prematurely at her negative conclusions. Observe that we did not affirm that these negative conclusions were scientifically erroneous. To have said that would have been to fall into the very error we were trying to avoid. We only said that they had been arrived at prematurely . . .[1]

Nowadays, when the believers in mind–body unity are in the majority and feel that they have won the battle—so much so that, on being asked recently to review a serious book on the evidence for survival after death, a well-known critic replied that his readers would not be interested—it is hard to realize how high feelings on the subject were running on both sides at the time. It is curious, incidentally, that with so many accounts of apparent ESP in the Scriptures, the Church was still too convinced of its diabolical nature to see a possible ally in psychical research.

Sniped at, then, from both sides, psychical researchers may be said to have shown some courage in carrying on in their mental no-man's-land, especially as many of them had considerable reputations to lose as scholars and scientists along other lines. Moreover, they themselves were only too well aware of the dilemma they had to face: that the facts disclosed by their empirical observations did not fit into any scheme of things known either to science or to common sense. One scientifically trained member of the S.P.R. remarked that if paranormal phenomena did occur he felt as if he must contemplate the possibility that the statue on the Albert Memorial might one day drop in to tea.

While studying the apparent evidence for survival after death the early psychical researchers anticipated Freud's revelations in the realm of psychology about the complicated subconscious life of man, and their discoveries emphasized the immense difficulty of obtaining evidence for survival which could not be given another explanation. They found, for instance, that at these uncharted levels of the psyche man is not only a great dramatizer; he is also extremely suggestible and seldom, if ever, forgets. Hence, in all honesty at the conscious level, a sensitive may serve up as coming from the discarnate information he or she had obtained years before by normal means—through overhearing a few words in a foreign tongue, perhaps, or glancing at someone

[1] Presidential Address, 1888, *Proceedings S.P.R.*, Vol. V, p. 271.

else's evening paper—and then, on the surface, had quite forgotten. Further, researchers came to realize that neither sensitive nor investigator could be quite sure of the source of information acquired by ESP: communications purporting to come from the dead could also be explained as subconscious dramatizations by the sensitive of information culled telepathically from the living. Even apparitions could be explained as hallucinations, created by the percipient himself to bring to his surface consciousness information telepathically received at a deeper level, which for some reason, possibly a biologically useful censorship, was prevented from emerging directly. On one occasion, for instance, a lady 'read' in a letter information which it did not contain, but which was afterwards found to be correct.

Researchers also discovered that a telepathic 'signal' can apparently lie in wait beyond the threshold of consciousness until the percipient's mood and situation enable it to emerge—until, say, he returns home from a cocktail party to sit in peace by the fire. Hence an apparition which appears to be of a dead man may in fact be a delayed telepathic 'signal' from one about to die. Further complications were the discoveries that the percipient *reacts* to the agent's situation, rather than *reproduces* it, that the emergence of an ESP 'signal' to consciousness may not be easy, or complete, or exact, and that ESP-type experiences seldom correspond in detail with related events. More often they take an oblique or symbolic form, much as emotionally charged material may do in a dream. They can also range from simple impulses towards apparently pointless action to vivid interior images or full-blown exterior hallucinations. And, to make things easier, they can even get mixed up with irrelevant material deriving from the percipient's own subconscious self.

It was eventually realized that, even were survival a fact, these complexities and ambiguities made watertight evidence of it almost impossible. Respect is due to the psychical researchers for frankly accepting this disappointment since their avowed hope had been to establish it without question by scientific means. The complexities and ambiguities also caused many probably genuine examples of mundane ESP to be discounted, for here, too, researchers felt that the collection of watertight evidence was the only way to justify their claim to a scientific approach; they were in no mood to take much interest in psychological complications for their own sake. Professor Sir Cyril Burt has pointed out that this attitude had its drawbacks. 'By placing the emphasis,' he writes, 'on the paranormal character of the phenomena, instead of on their psychological character, students of psychical research have to some extent impeded the scientific study of the problems involved and delayed acceptance of results already achieved.'[2]

Psychical researchers were, in fact, between the Devil and the deep sea, and in the end, particularly as behaviourism came to be orthodox psychological doctrine and mind and consciousness dirty words, they came to accept that only ESP produced to order under stringently controlled conditions would ever put it officially on the map. In consequence they concentrated their interest almost entirely on experimental work. From the beginning, of course, they had recognized this work as of vital importance, and, quite early on, some striking experiments had been done in which one person reproduced with surprising accuracy a number of target drawings made by another. But both persons were in the same room, so these hits could be dismissed by sceptics as due to fraud or to a lucky combination of unconscious whispering and hyper-acute hearing. Later on two members of the S.P.R. achieved some successes with target drawings at a distance, but the targets were not chosen at random, so it could be argued that, the experimenters being friends and having, presumably, similar interests, they could independently have chosen similar objects.[3]

Between 1915 and 1929 Professor Gilbert Murray made history with a long series of telepathic experiments about which Freud wrote to his friend, Dr. Ernest Jones, 'I confess that the impression made by these reports was so strong that . . . I should be prepared to lend the support of psycho-analysis to the matter of telepathy.' This so much unnerved Dr. Jones, who felt convinced that acceptance of telepathy would be fatal to psychoanalysis, that he sent out a circular letter describing its dangers. And from that point of view he may have been right, since the mental climate was such that one explanation put forward for Professor Murray's successes was that he had cheated!

[2] 'Physicality and Psi', *J. Parapsychol.*, Vol. XXV, 1961 (i), p. 31.
[3] Similar experiments have been done, with a considerable degree of success by Dr. Carl Bruck in Germany, René Warcollier in France, and Upton Sinclair in America. And in England, Whately Carington did a number of successful series with randomly chosen targets and widely scattered percipients.

Eventually, in the early 1930s, a partial solution to the problem of repeatable ESP was worked out by Professor J. B. Rhine of Duke University, North Carolina. Its principle was simple: one person, under supervision, would look through a pack of cards at a given time and speed while at the same time and speed another person in a different place would 'guess' its order. The advantage of this procedure was that, even if every 'guess' were not correct, the number of hits in proportion to the number to be expected by chance alone could be assessed statistically. (The assessments were made according to a method devised by Sir Ronald Fisher.) Thousands of these monotonous experiments were carried out in a number of places, many by Rhine and his colleagues, and many more with equal success by an English mathematician, Dr. S. G. Soal, of London University. They were an ironically prosaic conclusion to the pursuit of the erstwhile supernatural, but their results were dramatic enough, for over and over again a few 'guessers' in both England and America made an astronomically significant proportion of hits.

By the 1950s variants of this type of experiment had been going on for over twenty years, and the evidence for ESP provided by them had become so strong that such men as Professors Sir Alister Hardy and C. D. Broad stated in public that no one who had studied it dispassionately could reject it. But ardent sceptics were still not converted and one of them made a last ditch stand by suggesting that the nature of randomness itself had not been understood, hence statistics were not a reliable method of assessment. In 1957 Professor H. J. Eysenck put the absurdity of the situation in a nutshell.

> Unless [he wrote] there is a gigantic conspiracy involving some thirty University Departments all over the world, and several hundred highly respectable scientists in other fields, many of them originally hostile to the claims of the psychical researchers, the only conclusion an unbiased observer can come to is that there does exist a small number of people who obtain knowledge existing in other people's minds or in the outer world by some means as yet unknown to science.[4]

On the whole Professor Eysenck's psychological colleagues seemed more hostile to the idea of ESP than contemporary

[4] *Sense and Nonsense in Psychology* (Harmondsworth: Pelican Books 1957), p. 131.

physicists, partly, perhaps, because no suitable 'transmitters' or 'receivers' had been located in the human brain, so the idea offended against the current orthodoxy of behaviourism. (The phrase extra-sensory may also be to blame since it suggests that no such instrument will ever be found. A name for the phenomenon implying no more than that communication takes place by some means other than the *known* senses might have aroused fewer hostile reactions.) In the 1950s one American psychologist replied to a questionnaire on ESP: 'What are the quirks that lead an otherwise sensible psychologist to waste his time on so unrewarding a field of inquiry?' Another wrote in the journal, *Science*, that 'not a thousand experiments with ten million trials and by a hundred separate investigators' could make him accept ESP. And the Professor of Psychology at McGill University, Dr. D. O. Hebb, also wrote that he could not accept it, strong though the evidence was, because the idea did not make sense. The pattern is a familiar one, though in the past, when Copernicus expected to be hissed out of court for his preposterous suggestion that the Earth went round the Sun and Darwin said that to confess to a disbelief in the immutability of species was like confessing to a murder, outraged orthodoxy was on the other side of the fence. 'The mind,' said Wilfred Trotter, 'likes a new idea as little as the body likes a strange protein.'

In the 1950s, then, the situation looked like a stalemate. ESP still could not be harnessed to order, and merely to demonstrate its overwhelming probability did not satisfy critics who hated and subconsciously feared its heretical implications and past association with the supernatural. But now, in the sixties, a softer wind is beginning to blow. One reason for this is practical: the idea that telepathy might be a means of communication in space flight. In 1963 Dr. Eugene Konecci, Director of Biotechnology and Human Research in NASA (National Aeronautics and Space Administration of the U.S.A.), made the following remarks in a speech at the International Astronautical Congress in Paris:

> A concentrated effort towards a highly interesting problem in modern science—the nature and essence of certain phenomena of electro-magnetic communication between living organisms—is reportedly being pursued with top priority under the Soviet manned space program.
> Until recently these phenomena have in general been ignored by

Western scientists; however the many hypotheses involved are now receiving attention in world literature.

Specific U.S. experiments in energy transfer phenomena, or the relationship between the physical fields of particles and the non-demonstrable 'personal' psi-plasma field, are being carried out or planned under various advanced concepts.

... to Western scientists and engineers the results of valid experimentation in energy transfer could lead to new communications media and advanced emergency techniques, as well as to bio-cybernetical aids for integrating with a conceptual design of an ultimate operational flight system.

Such a design could result from a present NASA study on data subsystems and certain astronaut self-contained sensor systems. This vitally important study involves the function of the psychological information acquisition, processing and control systems.

In layman's language Dr. Konecci was talking about experimental attempts to achieve telepathy to order, but it is hardly surprising that he steered clear of the word itself, both in view of his own admission that 'no known rational explanation has yet been formulated as to the *modus operandi* of these unique means of gaining information', and of its association with the heresies of past psychical research about the possible dual nature of man. Just how embarrassing are these associations to contemporary orthodoxy comes out in some remarks made in 1964 by the late Dr. Norbert Wiener, on future trends in psychology.

> Many other considerations [he said] which up to the present have been situated in a somewhat shameful background, such as the study of direct communication at a distance, possibly by some sort of radiative phenomenon, are going to be subjected to a real trend in scientific examination, which will not be corrupted by the unscientific assumption that we are dealing with phenomena with no physical correlates.

All this is a step forward. It implies at least some breaking down of the curious but familiar dictum, that because we do not know how a thing works and do not like certain hypotheses about it, we must pretend that it does not work at all. (That dictum may be the other side of the penny about which Professor Michael Polanyi remarked recently, 'It is considered quite in order to teach absurd views that we do not believe, because we think they are scientific.')

NOTES ON CHANGING MENTAL CLIMATES

It may perhaps be assumed that if NASA and its Russian equivalent do find a method of achieving ESP to order, they will exercise a certain discretion about publishing it to the world. Still, in 1962 L. L. Vasiliev, Professor of Physiology at Leningrad University, published two reports of apparently successful experimental work done by himself and his colleagues, in one of which he quotes a distinguished Soviet rocket pioneer as saying that 'the phenomena of telepathy can no longer be called into question'. Vasiliev's second report, published in English under the title, *Experiments in Mental Suggestion*[5] records several series of experiments in which hypnotized subjects were sent to sleep and awakened at a distance, and at will. On occasion, the distance was as great as from Leningrad to Sevastopol. In some experiments, to eliminate the type of electro-magnetic radiations from the brain which Vasiliev and his colleagues had previously assumed to convey the signal, both hypnotist and subject were placed in metal (Faraday) cages, specially designed to obstruct these radiations; but the subject's responses, he says, were not affected. Although this result appeared to contravene the tenets of orthodox materialism, it did not disturb Professor Vasiliev, perhaps because he felt himself in less danger than his Western colleagues of seeming to be tainted by the heresies of the past. 'In every case,' he commented peacefully, 'the problem resolves itself into the search for a form of energy the properties of which are as yet unknown and specific to substances of such advanced nature as the human brain.'

In Czechoslovakia there has been a similar development. Dr. Milan Ryzl of Prague has trained a man who had hitherto shown no gift for telepathy to exercise it under hypnosis, and this man has passed increasingly difficult tests by various foreign investigators, including British and American psychologists. Here again is a hint that some unknown inter-personal factor may be at work, because so far no one in the West seems to have been able, like Ryzl, to train a subject who had previously shown no gift for ESP, to develop it under hypnosis.

In the realm of psychology, too, there are signs of at least a faint breeze of change, fanned in part by a few courageous analysts who have recorded evidence for telepathy between themselves and their patients. And they are beginning to ask the crucial question

[5] Church Crookham, Hants: Galley Hill Press, 1963.

about any ESP-type experience: Has this a psychological meaning? Does it make psychological sense? C. J. Jung, in particular, in the last few years of his life, flung two startling ideas into the pool of orthodoxy. The first was the principle he called synchronicity—that events can, so to speak, bunch together in time and space, not because one is *causing* the other but because their *meanings* are linked. The second was his insistence in his autobiography, published in 1963, that ESP had been a vital and recurrent factor in his own life and in his relations with his patients.

One key to this beginning of a change in the mental climate in relation to ESP may perhaps lie in the astonishing revolution in physics which is now going on, and which has already shown that the classical notions about the nature of time, space, and matter are not all-embracing when it comes to the realms of the very large and the very small. And now new research is revealing that in the realm of the very small the brain functions as a chemico-electrical instrument of surpassing delicacy and in a manner unsuspected in the days when it was ruled out as being in any way concerned with ESP.

The revolution in physics, of course, has been going on for over half a century, but new ideas need a lot of time to seep through and modify the general outlook, and the most fundamental change of all—to a concept of 'solid' matter as being convertible into elusive 'intangible' energy—is only now beginning to exert its influence in the outside world. In the early days of psychical research men could say that no mystery remained in the universe and that the subject of physics was exhausted. Now a pioneer of the new physics can write that 'the physics of the future a few centuries hence could well be as different from the physics of today as the latter is from the physics of Aristotle', and, again, that 'we must never forget how limited our knowledge must always be'.[6]

In such a mental climate ESP is coming to seem less of an 'egregious logical sin', and psychical researchers are more able to continue their empirical observations undeterred by the dictum that what they observe has no business to be there.

[6] Louis de Broglie, *Nouvelles Littéraires*, 2 March 1950.

V

PSYCHOLOGY AND PARAPSYCHOLOGY

Cyril Burtt

'Ἀθάνατος ψυχή, κοὐ χρῆμα σόν, ἀλλὰ προνοίας,
σοὶ δὲ τί τῶνδ' ὄφελος, ὃ ποτ' οὐκέτ' ἐὼν τότε δόξεις;
ἢ τί μετὰ ζῳοῖσιν ἐὼν περὶ τῶνδ' ματεύεις;[1]

1 THE PSYCHOLOGY OF THE SCEPTIC

Current attitudes to ESP

Dr. Rhine, in his recent review of the history of parapsychology, has described 'the towering rage of denunciation' which greeted the announcement that a laboratory for psychical research had been established at Duke University, and that strong experimental evidence had already been obtained in support of extra-sensory perception. 'Most surprising,' he adds, 'was the vigour of the hostile reaction from psychologists.'[2] In this country, Dr.

---

[1] 'The soul is immortal; yet that is no affair of thine, but of Providence. For thee—what profit is there in these matters? Some day, when no more, thou shalt believe it. Why then, while still among the living, dost thou search out such mysteries?' (Philostratus, *Life of Apollonius of Tyana*, the words of Apollonius when after death he appeared to a youth in a dream).

[2] J. B. Rhine, *Parapsychology: From Duke to the Foundation for Research on the Nature of Man* (Durham, N.C.: Parapsychology Press, 1965, pp. 14 ff.) In British academic circles the opposition has been neither so strong nor so sustained. Thirty years ago, when it was first proposed to recognize problems in parapsychology as permissible research-topics for higher degrees, there were a few who felt it their duty to nip this sort of thing in the bud; but today most of my colleagues, whether sceptical or sympathetic, would be perfectly willing to countenance, and even to assist, any well-planned scientific inquiry in this field.

J. L. Martin, in a 'survey of beliefs among educated males', similarly found that 73 per cent of those she questioned were

> convinced that genuine telepathy and clairvoyance were (as some of them put it) 'just old wives' tales'. The sharpest opposition [she adds] came from teachers and students of science; and the derisive assurance with which most of them pooh-poohed all such notions reminds one of the stories of Bishop Wilberforce's reception of the evolutionary origin of species: 'Mr. Darwin,' said his informant, 'has spent twenty years trying to prove his theory is true.' 'A person of intelligence,' retorted the Bishop, 'needs only ten minutes' reflection to see that the theory is utterly impossible.'

Accordingly, my purpose in this paper will be to examine the reasons why so many intelligent people still find it difficult—I will not say to accept, but even to consider—the conclusions reached by parapsychologists, despite the fact that they are based on what purports to be rigorous scientific research. The problem, as we shall see, is itself very largely a psychological question—a matter of half-conscious motives and implicit presumptions quite as much as of detailed knowledge or explicit reasoning. We shall therefore be concerned, not so much with the evidence adduced by those who defend these unorthodox views but rather with the types of argument advanced by their critics and opponents. Of these Professor Hansel is the most explicit and Dr. T. R. Willis is perhaps the clearest and the most concise:[3]

> We may [writes Dr. Willis] put the main argument in a nutshell. The conclusions of modern science are reached by strict logical proof, based on the cumulative results of numerous *ad hoc* observations and experiments reported in reputable scientific journals and confirmed by other scientific investigators: then, and only then, can they be regarded as certain and decisively demonstrated. Once they have been finally established, any conjecture that conflicts with them, as all forms of so-called 'extra-sensory perception' plainly must, can be confidently dismissed without more ado.[3]

[3] Shortly after the war, a series of discussions on psychical research, opened by Professor Broad and Dr. Soal, was arranged at University College, London, and included papers or talks given by sceptics and critics, such as Professor J. B. S. Haldane, Professor Hogben, Dr. Willis, Dr. Rackham, and others. Some of the papers, slightly abridged, have since been published in recent issues of the *Brit. J. statist. Psychol.* (Vol. XIII, 1960, and subsequent numbers), starting with an attack by Professor C. E. M. Hansel. The discussions and the correspondence in the scientific and popular Press which

But were this all, anyone who took the trouble to glance through the 'reputable journals' on parapsychology ought surely to accept such well-attested phenomena as telepathy and precognition, even if he feels doubts about the rest; for the amount of observational, experimental, and statistical evidence there recorded is quite as extensive as that which could be cited in defence of many of the more familiar scientific theories, which are now accepted without question. It is therefore, not so much the lack of evidence which is responsible for this summary rejection, but rather the apparent conflict between the supposedly 'supernatural' or 'paranormal' assumptions of parapsychology and what Professor Hansel calls 'the normal and perfectly natural framework of present day science'.

The fault lies, so we are told, with the logic and the methods of 'extra-sensory pseudo-science'. However, the popular conception of scientific methodology which both Professor Hansel and Dr. Willis expound seems grossly oversimplified. In the 'natural' (i.e. empirical or inductive) sciences the conclusions are never 'certain' or 'decisively demonstrated': they are only more or less probable, and are constantly liable to drastic revision. The kind of inference used is never a matter of 'proof' in the strict sense of the term; it proceeds by tentatively weighing the relative probabilities of two or more alternative hypotheses; and unfortunately in the ordinary scientific textbooks we rarely hear anything about the rejected hypotheses. Formally analysed, the mode of reasoning employed involves two distinguishable stages, and is based on two very different types of premiss; and once again the text book mentions only a single type—what Dr. Willis calls 'the results of *ad hoc* observations and experiments'.

The readiness with which we are willing to assent to any specific conclusion is always the result of two sets of factors: first, what most of us would describe as tacit preconceptions or presumptions (or as prejudices when we disagree with them); secondly, the trustworthiness of the first-hand evidence adduced in support of the conclusion proposed and the extent to which that evidence appears to exclude alternative interpretations, e.g. the possibility

---

followed, suggest that their criticisms reflect the views and attitudes of a vast number, if not the majority, of ordinary people at the present day. Professor Hansel's arguments, slightly modified, have now been set out at length in his book on *ESP: A Scientific Evaluation* (New York: Charles Scribner, 1966).

of sheer coincidence. Thus, to borrow the more precise terminology of the logician, we may say that the strength of any belief depends (i) on its 'antecedent' or 'prior probability', that is, the probability of the proposed hypothesis as judged by what we already know *before* any *ad hoc* observations have been made (and still more perhaps by what we do not know but automatically assume as a result of our half-unconscious wishes, hopes, and general beliefs), and (ii) on the 'likelihood' that the results observed would follow from this, that, rather than from any other particular hypothesis.[4]

Antecedent probabilities and their influence

Of these two factors the first, though all too easily overlooked, has frequently been the most influential in the older sciences as in the new. Consider the violent antagonism encountered by the theories of Copernicus and Galileo in astronomy, Buffon and Hutton in geology, Darwin and Huxley in biology—most of them theories which are now almost unanimously accepted. In these cases the resistance, as has so often been remarked, arose largely out of time-honoured metaphysical preconceptions or prejudices associated with religious beliefs. Today in his anxiety to avoid or

[4] Professor Hansel's treatment of the probabilities differs widely from my own. He takes each of the chief 'conclusive' researches separately, and argues that, 'if there is the *smallest possibility* of the successes undoubtedly achieved being explained by some *well established* process', then this explanation, 'however unlikely', should be preferred; each of them *could* be explained by joint trickery on the part of the experimenters concerned: hence the hypothesis of ESP can be unhesitatingly dismissed. This interpretation, it will be seen, implies a vast conspiracy, or set of almost simultaneous conspiracies, in Britain, France, Holland, and the U.S.A. Such a conspiracy on a world-wide scale seems reminiscent rather of a James Bond plot than an impartial weighing of evidence. Professor Hansel speaks in terms of an all-or-nothing verdict. No parapsychologists reported their researches as 'conclusive'. They would admit a 'small possibility' of trickery, but they would equally insist that it was highly improbable. Professor Hansel's book is certainly the fullest statement up to date of the case against parapsychology; and, if this is the best the critics can do, parapsychology would seem to be in a fairly strong position!

For the treatment of the probabilities adopted above, see C. D. Broad, *Religion, Philosophy, and Psychical Research*, (London: Routledge, 1953), p. 28, or, for a more general account, W. S. Jevons, *The Principles of Science* (London: Constable, 1900) Ch. XII, 'The Inverse or Inductive Method.'

disclaim any bias of this kind, the twentieth-century scientist tends to rush to the opposite extreme. This is particularly the case in discussions on psychology and parapsychology. The dominant schools of the present time—the behaviourists, the physicalists and the logical positivists—unite in assuring us that we can eliminate all bias due to our antecedent prepossessions, and thus view the universe with complete impartiality, only if we are willing to banish at the very start any concept or hypothesis which, in its origin or essential nature, is 'metaphysical' or 'theological'.

The popular version of this argument is perhaps most clearly stated by Dr. Rackham:

> It is [he says] a matter of history that traditional psychology—the 'logy' of the immaterial 'psyche' or soul—had its origin in the attempts of patristic and mediaeval theologians, like St. Augustine and Thomas Aquinas, to rationalize the myths and superstitions of the early Christian church. The remarkable advances of physics, physiology, biology, and biblical criticism during the nineteenth century at length freed the minds of bolder scholars and scientists from the needless and misleading dogmas which had hampered their predecessors. Nevertheless, many members of the educated public were unable to tolerate the loss of the faith in which they and their forefathers had been brought up. As a substitute therefore they sought refuge and consolation in the novel cult of spiritualism —a cult which, as Dr. Thouless has observed, was almost unknown in Western society until the latter half of the 19th century, the time when more orthodox creeds were losing their grip. It was a cult which strove pathetically to preserve and weave together all the old traditional beliefs in supernatural agencies, and, after dressing them up afresh in a pseudo-scientific jargon and tagging on a profusion of questionable experimental data, presented them as the latest offshoot of the new science of psychology. However [he observes in an ironical aside], whereas the ancient prophets and wonder-workers were depicted as men of holy character and spiritual gifts (though we should not forget the witch of Endor), our modern soothsayers are merely dubious 'mediums', usually paid for their efforts.[5]

[5] A Roman Catholic lady of my acquaintance tells me that, when as a girl she grew temporarily sceptical about her parental faith, instead of praying to St. Anthony to help her recover lost objects (as she previously had done), she had recourse to a medium: 'the medium', says my friend, 'was nearly, though not quite, as successful as St. Anthony—*teste diario*'.

Much the same view of scientific history has been painted by Professor Kantor in his *Scientific Evolution of Psychology* (1963), and popularized by Professor Hansel, Dr. Taylor, Dr. McLeish, and other contemporary behaviourists. Like Rackham, Kantor, drawing heavily on Gibbon, ascribes 'the decline and fall of Graeco-Roman naturalistic science' to 'the official adoption of Judaeo-Christian supernaturalism by Constantine and his successors'. 'These dogmas,' we are told, 'have continued to bedevil the progress of psychology until their last feeble flicker in the teachings of James and McDougall.' But 'even then, and even after Watson in America and Ryle in Britain had finally exorcized the "ghost in the machine", it still seems to haunt the disciples of James and McDougall at Duke and elsewhere'.

Now it must be frankly confessed that many parapsychologists, by the very phraseology they use, appear at times to play straight into the hands of such critics. Dr. Rhine, for example, announces that 'the claims of psychical research are the very substance of religious belief'; indeed, as Professor Flew declares, the style in which writers on such topics report their findings, instead of suggesting a cautious impartiality, 'often generates an atmosphere of uplift-cum-mystification that tends rather to intensify the suspicion with which the hard-headed sceptic naturally views such revolutionary claims'.[6] Yet the note of alarm that so often sounds in the critics' indignant protests suggests that they, too, are swayed by emotional influences as much as by rational. Indeed, at times one is reminded of the reply of the village girl in Conan Doyle's story: 'We don't believe in ghosts, but we'm horribly afraid of them.'

Arguments and objections, like those I have quoted, so constantly crop up in current controversies about psychical research that it seems essential at the outset to point out the fallacies and inaccuracies which they contain. In the first place, even if the facts alleged were correctly stated, they would by no means disprove the views or the theories reached by the modern parapsychologist. Practically every branch of science had its origin in some type of

[6] J. B. Rhine, *New Frontiers of the Mind* (London: Hutchinson, 1950), p. 43; A. G. N. Flew, *A New Approach to Psychical Research* (London: C. A. Watts, 1953), p. 113. This criticism, however, plainly refers rather to Dr. Rhines's popular expositions than the articles in which he reports his experimental and statistical investigations; and I for one would hold that he is justified.

superstition, priestly or otherwise—astronomy in astrology, chemistry in alchemy, physiology and medicine in the doctrines taught in the temples of Aesculapius. Nor is it the purpose of psychical research to advocate any specific doctrines—spiritualist or otherwise [7]—but merely to apply modern scientific methods of research to what James has called 'residual phenomena'—that is, the anomalous and sporadic observations and reports which have given rise to the 'speculations' and 'superstitions' to which the critics refer. But secondly, the notion that Graeco-Roman science and philosophy were distinguished by a strict 'materialistic realism' (Kantor's phrase) and that 'the concept of an immaterial and immortal soul is a relic of Judaeo-Christian supernaturalism handed down from the Dark and the Middle Ages'—all this is a travesty of philosophic history. [8] And the same holds good of Dr. Rackham's sweeping declaration, that 'the common sense of

[7] See the discussion of spiritualistic cults in the admirable Pelican book by Dr. D. J. West, formerly Research Officer for the Society for Psychical Research (*Psychical Research Today*, 1962).

[8] The idea of an immortal soul, distinct and separate from the material body, is Greek, not Jewish or Christian, and was repeatedly condemned as heretical. The 'Apostles' Creed' insists on the 'resurrection of the *body*'. An impartial reader can hardly help being startled by the strident tones which appear in the modern critic's voice when he inveighs against 'theological dogmas like that of "the soul" or "personal immortality" creeping into what psychologists hand out as "science"'. One seems to be back in the eighteenth century, listening to the cries of 'Écrasez l'infâme'. 'Anti-theologians', such as Watson, Kantor, and their younger disciples, forget how much the revival of natural science after the Dark Ages owed to the observations, experiments, and theories of men who were themselves monks, priests, or pilgrims—Jordanus, Albertus Magnus, Petrus Peregrinus, and, above all, perhaps Grosseteste and his school in Britain. If everything that had its origin in 'preconceptions hailing from the Judaeo-Christian religion' were expunged from modern science (including psychology), little would be left. The authentic historians of science and philosophy (Whitehead, Collingwood, and Crombie, for example) continually remind us that it was these very preconceptions that provided 'the basic assumptions of natural science—that the whole universe is a unity, that all its processes are subject to laws, and that these laws are intelligible to human reason'—postulates which reached their climax in the unified scheme of Newton, who himself emphasized their theological origin (see the general scholium to Bk. III of the *Principia* and his letters to Bentley). 'Where there was no monotheism, with its notion of a single supreme Power, respecting law and reason, there [says Collingwood] was no science of nature' (*An Essay in Metaphysics*, London: Oxford University Press, 1940, pp. 201 ff.).

almost every pagan philosopher—from Xenophanes "the first rationalist" until the Athenian schools were finally closed a thousand years later—would have scorned the stories and reports that fill our journals of parapsychology as figments of popular superstition on a par with the myths and miracles of the Homeric heroes'.

Let me give just one example out of many which these historians of parapsychology seem to have overlooked. It will at the same time help us to picture for ourselves the types of phenomena with which we shall be concerned, and to realize the ubiquity of the beliefs that we are about to examine.

In the first century of our era the most celebrated philosopher of the day, rivalling the disingenuous Seneca in fame, was Apollonius of Tyana, leader of the Neopythagorean school, and a thinker who exercised considerable influence on the later Neoplatonists and even on Christian writers. Unfortunately, only fragments of his numerous works have survived. However, an Assyrian disciple who accompanied him on his wanderings left memoirs in uncouth Greek which fell into the hands of the Empress Julia Donna.[9] She requested Philostratus, a member of her salon, to rewrite the biography. This he did, incorporating data from other contemporaries and from what he himself learnt while visiting the place where Apollonius had taught.

Apollonius was born of a wealthy family about 4 B.C. He studied philosophy, and renounced marriage, meat, and wine. Taking up his abode first of all in the temple of Aesculapius (god of medicine), he acquired such a reputation for sanctity and spiritual gifts that the sick flocked to him to be cured. He cast out demons, raised the daughter of the consul from the dead, was imprisoned by Nero but miraculously escaped, and became the friend of the Emperors Vespasian and Titus. He gave all his goods to the poor, and urged the rich to do the same. He travelled widely, endeavouring to purify the religious cults wherever he went; he discussed 'the world-soul' with the Brahmins in India, and there witnessed

[9] Julia Donna was the wife of the Emperor Septimus Severus who died at York (A.D. 211). She was keenly interested in science, philosophy, and letters; her circle included Galen the physician, Ulpian the lawyer, and Diogenes Laertius (author of the *Lives of Philosophers*)—men by no means devoid of a critical outlook. Philostratus' *Life of Apollonius*, in eight 'books', has survived intact.

human levitation; he set out to discover the sources of the Nile, which he located in the cataracts. He could read the thoughts of others, paralyse a man with a glance or a touch, foretell the future (for example, the death of Titus), and possessed the power of second sight (as in his vision of the murder of Domitian). His disciple has left no record of his death. But it was said that he lived to the age of 100, and finally entered a temple, where he 'ascended into heaven'. Many years later he appeared in a vision to a young student of philosophy, assuring him (in the Greek hexameters quoted above on p. 61 that the soul was indeed immortal.

Here we have all the typical paranormal phenomena investigated by modern psychical research—telepathy, precognition, clairvoyance, psychokinesis, hypnotic suggestion, apparitions, and communications from the dead. But what is of special interest are the explanations hinted at by Apollonius himself, partly in the *Apologia* which he wrote in reply to critics, and partly in his letters and other writings. Both the claims and the interpretations proposed are strikingly analogous to those put forward by Jung in his recent autobiography. As it happens, we also possess a long letter written by Eusebius, the Christian historian,[10] criticizing a later biography of Apollonius. Eusebius says that he is prepared to accept most of the facts recorded by the earlier and contemporary writers, with some allowance for the usual tendency of devoted disciples to embellish or exaggerate. As Apollonius himself had intimated, he says, what looked like miracles to the naïve crowd might easily have had a natural explanation: a shrewd and experienced traveller could often have been able to guess thoughts of others or forecast the future; and, being trained in medicine, he might have 'detected some spark of life in the girl whom he revived, which her own physician had not noticed'. But supernatural interpretations Eusebius refused

[10] Eusebius was bishop of Caesarea and a friend of Constantine. In the controversies of Nicaea he figures as the leader of the moderate or liberal party, insisting, as Origen had done, that scriptural pronouncements on the mysteries of the universe should be treated as allegorical, poetical, or symbolic rather than as literal. The work criticized was a fanciful panegyric of Apollonius, written by Hierocles (a provincial governor under Diocletian and a contemporary of Eusebius), extolling his hero as a wonder-worker and demi-god, 'far more worthy of temples in his honour than the Jesus of the Christians'.

to admit. 'If asked my reasons,' he writes, 'what I would say is this: there are bounds to nature which prescribe and restrict what can happen in the universe—rules and limits that are imposed on all things; by these the entire mechanism and structure of the whole cosmos (μηχάνημά τε καὶ ἀρχιτεκτόνημα τοῦ παντὸς κόσμου) is continually being worked out, ordered, and determined according to unbreakable laws and indissoluble bonds' (*op. cit.*, c. VI). Let those who read Professor Kantor and Dr. Rackham take note that it is the pagan Greek who defends the 'supernatural hypothesis' and the Christian bishop who rejects it.[11]

Metaphysical presuppositions

However, what Professor Kantor, Dr. McLeish, and other critics reject as '*religious* concepts and pre-conceptions' are really antecedent *philosophical* assumptions; and, as is clear from the arguments they employ, most of their hostility to parapsychology springs, not so much from any critical scrutiny of the factual evidence, but rather from certain metaphysical presuppositions, with which the parapsychologist's conclusions would admittedly conflict. Now, as sciences, both psychology and parapsychology are intimately concerned with man and human problems. Hence, unlike other sciences, they are peculiarly apt to become entangled with the quasi-philosophical issues which, consciously or unconsciously, affect every critical thinker. Indeed the psychologist, who is often by profession an educationist or therapist, or himself engaged in training educationists and therapists, cannot afford to pass lightly over the deeper implications of his conclusions and his theories, or to ignore their impact on those whom he is educating, training, or treating. He is therefore, or should be, accustomed to pay special attention to the metaphysical or metapsychological assumptions on which his scientific inferences are based.[12]

[11] Had Eusebius's apostate pupil, the Emperor Julian, succeeded in reinstating 'the gods dethroned and deceased', then, as Swinburne hints in his poem, the 'occult sciences' taught by his 'mystical master', Iamblichus, would have gained a still firmer hold, and we should have heard still more clearly 'The murmur of spirits that sleep in the shadow of gods from afar'.

[12] By 'metaphysics' I understand that branch of philosophy which discusses the nature of 'reality' or 'being' in its most general aspects (Ewing). Since these terms are often misunderstood as implying some kind of absolute

PSYCHOLOGY AND PARAPSYCHOLOGY

Of the various types of metaphysical theory dualism is that which commends itself most readily to reflective common sense. I use the word dualism to cover any philosophic doctrine which holds that there are apparently two irreducible modes of being—usually termed mind and matter. To preserve an even balance between the two, however, is always difficult; and during the last century or so there has been a growing tendency to repudiate the 'bifurcation of nature', which all dualistic views entail, and to regard only one of the two modes of being as fundamental.[13] Monistic theories consequently fall apart into two contrasted groups—variously termed 'idealism' or 'mentalism', and 'empiricism', 'materialism', or 'physicalism', respectively. The empiricist contends that the sole source of knowledge is sense-experience, that reality is just an affair of 'blind brute fact' (fact being what is directly or indirectly verifiable through observation), and that what cannot be verified or falsified is meaningless. This attitude has commonly issued in a tough materialistic monism. It has been poignantly expressed by Russell in one of his earliest papers:

> That the world which science presents us is purposeless and void of meaning, that man's origin and growth, hopes and fears, are but the outcome of accidental collocations of atoms, that no heroism

[13] The general reader should be warned that the much publicized objections to dualistic doctrines are by no means so conclusive as is now commonly assumed: (for counter-criticisms, see, for example, A. O. Lovejoy, *The Revolt Against Dualism*, (London: Allen & Unwin, 1930).

---

reality, many would prefer to define it as 'the study of experience as a whole' (Mackenzie): the word 'experience' emphasizes the fact that the problem is as much epistemological as ontological. Metaphysics at the moment is under a cloud. Yet it is obvious that the various natural or physical sciences deal only with certain limited aspects of reality or experience; furthermore there are other ways of approaching reality besides those of science. Each of the special sciences begins by adopting, tacitly or explicitly, certain unproved assumptions and undefined concepts (space, time, space–time, substance, matter, minds, laws of causation, uniformity, conservation, etc.), which are clearly metaphysical, and ends with certain empirical, but necessarily provisional, conclusions. Hence the need for a more comprehensive theoretical and critical study which goes beyond or behind physics, and considers what presuppositions are in fact tacitly assumed, how far they are justifiable, and in what ways they may be harmonized. Thus science deals with what we know; metaphysics with what we don't know, but feel obliged to postulate.

can preserve an individual life beyond the grave—all these things, if not quite beyond dispute, are yet so certain that no philosophy which rejects them can expect to stand; only within this scaffolding, only on this firm foundation of unyielding despair, can the soul's habitation henceforth be built.[14]

Nevertheless, as a metaphysical hypothesis, materialism has been rejected by the vast majority of metaphysicians; it has formed the creed of the scientist rather than of the philosopher. Moreover, as we shall see in a moment, since Russell penned these lines, materialism in the strict and literal sense has become no longer tenable, even as a scientific theory. Hence it would be better to call the present-day version of it 'physicalism'—a name I shall use to designate all those theories which regard reality as in principle capable of a complete and satisfactory description in terms of the current concepts of the physical sciences.

Idealism reverses the position. It points out that what we call matter can be known only through our consciousness of material things. We must therefore take consciousness as basic, and regard physical phenomena as no more than the delusive manifestations or appearances of an underlying reality which is intrinsically mental or spiritual. The idealist is thus led to the belief that the universe is not wholly indifferent to ideals or values, that goodness, rationality, and even beauty are in some sense objective, that consciousness is no otiose by-product, that 'life is not destitute of meaning, nor reality altogether unintelligible or unfriendly'. Time is usually held to be unreal. Most idealists therefore hold that minds (or at least mind) are in some sense conserved, though they generally incline (as Ward does) to some form of 'spiritual monism' rather than (as McTaggart does) to a theory of a 'plurality of self-subsistent selves'—a theory which, as he develops it, involves both pre-existence and reincarnation. They might also be expected to favour paranormal sources of knowledge, particularly telepathy, precognition, and communications from 'post-existent self-subsisting selves'. Actually, however, they rely mainly on rational deduction from *a priori* premisses, and tend to

[14] 'A Free Man's Worship', *The Independent Review*, December 1903 (slightly condensed). See also the eloquent passage, which figures in so many prose anthologies, in A. J. Balfour's *Foundations of Belief* (London: Longmans, 1895), pp. 33-4.

cast a scornful eye on any quest for empirical evidence—'putting the ear to a tombstone for echoes from a life that is past'.

In the case of those who are not professional metaphysicians the choice between one or other of these various standpoints seems usually to be decided, not on the basis of any kind of proof, but simply because the alternative selected appeals to them as more satisfying. Just as people differ in their preferences for different styles of art—classical or romantic, abstract or realistic—for reasons which seem to them self-evident and therefore neither to need nor to admit of any formal demonstration, so also they differ in the kind of world-view they feel themselves impelled to adopt. Despite their heroic efforts to rationalize it, the main motives for their choice are at bottom unconscious—the product of some deep-seated difference in temperament. Every psychologist remembers James' apt distinction between 'tender-' and 'tough-minded types'; 'the tender-minded type', he tells us, 'is a born idealist; the tough-minded a natural empiricist'.

Now it must, I think, be owned that many of the leaders in the movement for psychical research—Myers, Barrett, Sidgwick, and Lodge (though not Podmore) among the pioneers, Rhine and Pratt among contemporaries (though not Soal or West)—have exhibited a marked and indeed an avowed bias of the former type; and their tacit predilections may, as their critics allege, have sometimes coloured their observations and deductions. Perhaps too, as Dr. G. R. Price in America and Professor Hansel in this country have maintained, this frank and ready optimism may have encouraged the mediums and the subjects on whom they have relied to play up to their sanguine expectations; there may even have been a modicum of trickery, self-deception, or unconscious wish-fulfilment. Such things are not unknown in other branches of science.[15] Yet it is only fair to recognize that the parapsychologists I have named, together with their chief co-workers, have been fully alive to these irrelevant influences, and have endeavoured to guard against them by laying down the requisite 'canons of evidence', and by formulating conditions to which all parapsychological investigations should conform. On the other

[15] Mendel's remarkably accurate results (so we have lately been assured) must have been partly due to the data reported by his venial gardener, who guessed only too well the kind of figures his master expected. And everyone remembers the hoax of the Piltdown skull.

hand, we cannot altogether acquit the empirical critics of bias in the opposite direction—a bias which often distorts their versions both of the facts observed and of the methods employed.[16]

Critics belonging to what I have called the empirical school contend (in Dr. Hudson's words) that 'natural science—including psychology, if it is to take its place as a natural science—must renounce all metaphysical concepts, postulates, and arguments, which are by their nature unverifiable, and therefore, as Ayer and Carnap have shown, devoid of meaning'. This, however, is itself not a scientific, but a metaphysical pronouncement, and one which most scientists, as well as most philosophers, have now abandoned.[17] Despite his own disclaimers, Professor Hansel's initial assumptions are in fact patently metaphysical. He starts his 'refutation of all paranormal phenomena' by codifying what he considers to be the 'rules and basic requirements of all sound scientific theorizing'. They consist, he tells us, of two sets of *a priori* postulates restricting permissible scientific hypotheses to certain traditional types, and he summarily bans any and every deviation as scientific heresy. (i) The first set states that *all* events are

[16] Professor Skinner, for example, has declared that the experimental results obtained by Dr. Soal with his most successful telepathist (Basil Shackleton) were entirely vitiated by the fact that he relied on cards shuffled by hand to secure a random order: Dr. Soal has replied by referring to his original report, and pointing out that no cards were used and no shuffling of any kind was involved. Similarly, Professor Hansel's descriptions of Dr. Soal's latest research and of Dr. Pratt's earlier investigations imply such obvious misreadings of their published accounts that his criticisms have little relevance to their actual findings. Let me add that these and other writers too readily assume that all who consider post-mortem survival a genuine possibility are 'motivated by some personal desire or fear'. Both McDougall and Broad, have declared they would "prefer total extinction at death to venturing anew on another life" (W. McDougall. *Body and Mind*, London: Methuen, 1928, p. xiv; C. D. Broad, (*Lectures on Psychical Research*, London: Routledge, 1962, p. 430). And no one who has read Broad's essays on *Religion, Philosophy, and Psychical Research* (1953) can plausibly maintain that his views on parapsychology have been influenced by his 'acceptance of the traditional dogmas of religion'.

[17] Ayer no longer insists on the principle of verifiability as he originally stated it; as for the scientists, it will suffice to quote Margenau: 'Every scientist *must* invoke assumptions and rules of procedure, which are not dictated by sensory experience. To deny the presence, indeed the necessary presence, of metaphysical elements in any science is to be blind to the obvious' (*The Nature of Physical Reality*, New York: McGraw Hill, 1959, pp. 12 ff., [his italics]; in a later chapter he formulates what he holds to be the 'metaphysical requirements necessary in every empirical science').

causally dependent on preceding events in the world of space and time, and that the modes of causal dependence must be those enunciated in the 'laws of mechanics'; (ii) the second set imposes similar limitations on the ways in which material events and conscious events may be assumed to interact, and the means by which conscious experience or information can be transmitted from one person to another. It is evident that clairvoyance and precognition are incompatible with the former, and that telepathy, psychokinesis, and communications from the dead are excluded by the latter. So convinced is Professor Hansel of the universal validity of these restrictive postulates that he ends by declaring that 'the *a priori* arguments based on them . . . may even save time and effort in scrutinizing the [parapsychologists'] experiments. . . . In view of the *a priori* arguments against it we *know in advance* that telepathy, etc., cannot occur.'[18] All this surely is to copy the attitude of Galileo's 'religious opponents', who also claimed to 'know in advance' that the phenomena observed through the telescope 'could not occur', and therefore declined to look.

What, then, is the warrant for this confident claim to virtual omniscience?[19] Professor Hansel, and many of the other writers who subscribe to the same dogmatic postulates, apparently suppose that to any unprejudiced thinker each and all of them must

[18] C. E. M. Hansel, 'Experiments on Telepathy in Children', *Brit. J. statist. Psychol.*, Vol. XIII, pp. 175–8: cf. *idem, New Scientist*, Vol. V, 1959, pp. 459 ff. Professor Hansel's forthright attitude reminds one of Helmholtz's retort when Barrett, Crookes, and other F.R.S.s defended telepathy. 'Not the testimony of all the Fellows of the Royal Society, nor even the evidence of my own senses, would lead me to the belief in the transmission of thought independently of the recognized channels of sense'; or of Hebb's most recent pronouncement: 'My own criteria, derived from physics and physiology, declare that ESP cannot be a fact, despite the behavioural evidence that has been reported'; but he admits that this is 'in the literal sense, just prejudice'.

[19] Professor Broad, who had already formulated a very similar set of 'basic limiting principles' (*Religion, Philosophy, and Psychical Research*, pp. 9 ff.), says that 'some of them seem self-evident; others are supported by the empirical facts which fall within the range of ordinary experience and the scientific elaboration of it, including orthodox psychology'. He, however, concludes that 'the paranormal facts which have been established to the satisfaction of everyone who is familiar with the evidence and is not the victim of invincible prejudice . . . call for very radical changes in a number of our "basic limiting principles".'

be 'intuitively obvious', or, in Dr. Hudson's phrase, 'self-evident'. But, as the history of science all too plainly demonstrates, what seems self-evident to one man may seem a mere arbitrary assumption to another, and palpably false to a third.[20] What we are to take as our postulates or axioms is therefore always a matter of convenience or convention. Any body of knowledge—mathematics, physics, and all the other departments of science—can be plausibly organized into a variety of widely different systems. If in one system A and B are chosen as the initial definitions and axioms, C may follow as a derivative theorem and D may be demonstrably inconsistent with all three. With another set of definitions and axioms, both C and D may be formally deduced, and A perhaps shown to be only approximately true. In certain mathematical systems some expressions are literally 'irrational' and 'imaginary'; in others, such quantities are not only legitimate but also extremely useful. And, as Gödel has shown, in every system there are well-formed and meaningful propositions which are undecidable within that system. Furthermore, when an appropriate theoretical model has been selected or constructed there still remains the semantic problem of linking the symbols, terms, or abstract concepts of the system with the observable data of the experimenter's laboratory or of everyday experience; and the 'rules of correspondence' formulated for this purpose will again vary with different scientists and with different stages in the development of the same science.[21]

[20] It has not, I fancy, been sufficiently realized that what appears self-evident depends largely on each man's mode of thinking—particularly on the type of mental imagery he habitually uses. The visualizer sees the world as a concrete material structure visibly existing in a quasi-Euclidean space, and he pictures interaction in the form of visible movements of mechanical components or particles. The motile, i.e. the man whose imagery is mainly kinaesthetic, takes force or energy as the basic concept. A verbalizer, who (like myself) may be almost devoid of visual or kinaesthetic imagery, tends rather to think in terms of abstractions, and for him material and mechanical structures lose much of their theoretical importance.

[21] E.g. for me the term 'sensation' denotes something that I (the speaker) can observe by introspection; for behaviourists, like Dr. Taylor or Professor Hansel, it denotes a 'mode of behaviour' which he (the writer) can observe in someone else. What I have called 'rules of correspondence' are roughly equivalent to what other writers term 'epistemic correlations'; so-called 'ostensive definitions' and 'operational definitions' are perhaps the types most familiar to the student of psychology.

Most psychologists, however, have come to realize in the course of their own work that apparent self-evidence is so frequently deceptive that it cannot possibly be invoked as an infallible basis for sweeping generalizations about the concepts or assumptions which are, or are not, permissible in any particular branch of scientific inquiry. Instead, those who attempt any such formulation,[22] now commonly follow Mill and other empiricists, and maintain that what they take to be the 'universal laws of nature' or the 'limiting principles of science' are, as Dr. Clarke puts it, 'truths conclusively established by induction from what can actually be verified by observation, that is, by actual perception'.

What perception shows us [he continues] are always physical things at rest or in motion—not of course simply chairs and tables, clouds or motor-cars, but the pointers, the specimens, the scales, and so forth of the scientist's laboratory. For the scientist therefore 'reality' must be confined to the facts, concepts, and theories derived from, or reducible to, sensory experience: there is no other way to attain reliable knowledge of the universe in which we live. Hence the psychologist, like every other scientist, must limit himself to the descriptions of observable behaviour, that is, of material bodies or organisms whose movements we can see (or hear in the case of 'verbal behaviour'). It follows that the non-sensory or extra-sensory does not exist.

Professor Skinner's arguments follow much the same lines.

Dr. Clarke's arguments put into words the implicit views and attitudes of most of those who dismiss the conclusions of parapsychology as 'flying in the face of science and common sense', or, like the people interviewed by Dr. Martin, meet all such suggestions with a blunt rebuff. We must therefore examine the arguments a little more closely, and ask how far 'sensory experience' and the 'concepts and theories derived from or reducible to' such experience can reasonably claim the high validity ascribed to them.

---

[22] Cf. C. C. Pratt, *The Logic of Modern Psychology* (New York: Macmillan Co., 1939). The 'universal laws of nature' on which Dr. Clarke lays chief stress are 'first the conservation of matter which forbids the idea that, with the birth of each individual, a new unit of mental substance is created out of nothing, and secondly, the conservation of energy which forbids any kind of direct interaction between one mind and another (such as is implied by telepathy and the like) or direct interaction between mind and matter (such as is implied by clairvoyance or psychokinesis)'.

II THE LIMITATIONS OF SENSORY PROCESSES

Neither those who have adopted the fashionable phrase '*extra-sensory* perception' nor those who reject it have stopped to explain what other forms of perception they tacitly contrast with it, i.e. what exactly they understand by simple '*sensory* perception.'[23] Dr. Clarke and others who think with him apparently regard it as a synonym for 'observation'. Now 'observing', i.e. seeing, touching, and hearing, and the other types of 'sense-experience', are all forms of cognition: they are (to borrow Stout's description) modes of being 'directly aware'. Hence the proper question to ask is, not whether extra-sensory *perception* ever occurs (the very expression, I agree, is almost a contradiction in terms), but whether there is any such thing as extra-sensory *cognition*. The answer is undoubtedly 'Yes'. Pleasure is not itself a sense-datum nor is it a qualification solely of sensory experience. Logical relations can be directly cognized; but they, too, are not forms of sensory experience. Many psychologists would extend the list, and include such things as pseudo-perceptions based on processes equivalent to 'unconscious inference' (e.g. the visual perception of distances and depth), and a variety of intuitive apprehensions, such as meanings, hunches, insights, the sense of value and beauty, and even mystical experiences. Hence to reject any form of extra-sensory cognition merely because it *is* extra-sensory would be just a clumsy way of begging the question.

In any case, no psychologist who accepts the principle of human evolution could possibly subscribe to the view that 'reality for the scientist must be confined to concepts and theories derived from sensory experience'. Our sense-organs were evolved, not to enable us to 'attain a reliable knowledge of the universe', but merely to help our particular species to survive in a particular kind of environment. We live on a planet that is by no means typical. Just as fishes are adapted to life in the water, and birds to life in the air, so man is one of the many animals that have be-

[23] In *Science and Psychical Phenomena* (New Hyde Park, N.Y.: University Books, 1961), G. N. M. Tyrrell has a brief chapter on 'Problems of Sense Perception' (pp. 138–53), mainly with a view to showing that 'sense-perception does not put us in immediate touch with physical objects'; but his discussion hardly bears on the points I wish to bring forward here.

come adapted to life on dry land. Our senses, our joints and muscles, our posture and our general bodily structure have been modified and developed to cope with the problem of survival in these very specialized circumstances. The things that are of vital importance to us consist primarily of solid or nearly solid objects of medium size, ranging from a grain of dust to hills and cliffs. Most of them are things we can reach, lift, pull, push, chase, or tear to pieces with the aid of the muscles of our legs and of our hands and fingers: and, when they move, their speed differs little from our own. Our eyes, which have become our dominant sense-organs, respond only to the necessary but narrow range of radiations. Our well-developed brains enable us to devise and make things as well as to use things. Thus our species might more suitably be described as *homo opifex* than as *homo sapiens*. By our very nature we tend to look upon the world as a collection of material objects, located or moving in empty space: all are pictured or thought of in visuo-muscular terms. Hence the early development of the mechanical sciences, with their concepts of matter and mass, and hence the overriding importance for practical purposes of the law governing terrestrial gravitation. Had we been intelligent snipe-eels, living in the depths of the ocean, hydrodynamics (which defeated Newton) would doubtless have been our earliest science, and our earliest cosmological scheme would doubtless have been based on some form of field-theory: Euler's laws of fluid mechanics would have been discovered long before Newton's laws of motion. On Venus the laws of thermodynamics might have been developed first. There is a still more remarkable limitation. Since most of the solid objects we encounter on this planet carry no electrical charge and exhibit no magnetic properties, no sense-organs have been evolved for the reception of electricity or magnetism. Important as we now know them to be, both are, for human beings, 'extra-sensory properties'.

The twentieth century, however, has witnessed a profound revolution in the conception of the world as formulated by contemporary physics. The elementary constituents of which it is built up are no longer conceived after the fashion of visible bits of matter, rigorously obeying the 'laws of mechanics' like so many colliding billiard balls. They are, on the contrary, almost incredibly minute; the speed with which they interact is almost incredibly swift:[24] and their mode of interaction often flatly contradicts the laws of classical physics. Although the scientist's manipulations and observations have still to be made with human eyes and hands, eked out with ingenious instruments, the hypothetical entities and processes which he infers are beyond the range, not only of human sense-perception and the electron-microscope, but even of human imagination. Indeed, as Heisenberg has insisted, 'the very attempt to conjure up a picture and think of them in visual terms is wholly to misinterpret them'. We are thus faced with a proliferation of new and elusive concepts which have no counterparts in our everyday household world or even in the world of classical physics.[25]

[24] A few figures may serve to bring home the radical nature of these changes. For nineteenth-century physics, which Professor Hansel still takes for his 'scientific framework', the hypothetical atom was the smallest particle recognized. The radius of an atom is about one ten-millionth of a millimetre. When attempts were made to estimate the size of an electron it was calculated that its radius was less than one ten-millionth of the radius of an atom; but in point of fact, the electron really has no size, and the traditional concept of space breaks down when we are dealing with the problems of nuclear physics. The mass of the electron is only 9×10^{-28} gram (a ninety thousand million million million millionth part); consequently, when we are dealing with these submicroscopic phenomena, the 'Newtonian force of gravitation', which seems so important in everyday life and on which Dr. Taylor lays so much stress, is virtually negligible. On the other hand, the electric charge possessed by an electron ($1 \cdot 76 \times 10^7$ electro-magnetic units per gram) far exceeds that of any other object or entity, whether observable or hypothetical. If the electron obeyed the laws of classical physics, it would spiral into the nucleus, and the whole atom collapse; the protons would repel each other, and the nucleus itself would be dispersed. A fresh set of 'basic limiting principles' have therefore to be introduced, apparently undreamt of in Professor Hansel's philosophy. The speeds involved similarly outrun the scope of anything we can conceive. To make the simplest observation with our ordinary sense-organs generally requires a period of at least one second. But the natural period for so-called 'strong' nuclear interactions is only about 10^{-23} seconds. Thus a particle which has an average lifetime of 10^{-10} seconds (one thousand-millionth of a second) really has an amazingly long life. One further illustration may serve to bring out the paradoxical nature of such phenomena: if we attempt to describe the apparent behaviour of a single electron when fired at a thin screen of metal containing two minute holes, we should be constrained to infer that the particle passed through the screen in two places at once—a feat which has never yet (as far as I am aware) been performed by the ghosts of either folklore or psychical research.

[25] How remote the basic constructs of modern physics are from the observable contents of sensory experience is shown still more strikingly by the

So far, then, my main contention may be summed up as follows. The fundamental concepts of modern science are so remote from actual observation and from ordinary sense-perception that it becomes ludicrous to insist, or even to suggest, that they should be 'limited' by the 'basic principles' which were derived from the observable behaviour of what I have called man-sized objects and processes. Nor can we expect the generalizations drawn from these new fields of inquiry to enjoy the same kind of quasi-intuitive certainty accorded to classical physics of the Newtonian or mechanistic era. Year by year it becomes more and more obvious that physical science tells us far less about the universe in which we live than we have hitherto been prone to assume. The 'firm foundation of physical science', on which Professor Hansel and others base their arguments, turns out to be only one of a series of progressive stepping-stones, and by no means the last or surest.

---

history of 'energy' and Einstein's unexpected identification of it with 'matter' or 'mass'. In their origin 'force' and 'energy' ('capacity for work') were anthropomorphic or rather psychomorphic notions, derived from our own conative experience; and once again a psychologist may be permitted to suggest that the old distinction between matter and energy resulted from the way biological needs determined the evolution of our senses. Our tactile perception of the gravitational effects of mass (e.g. a grain of sand falling on the skin) requires a stimulus of at least $\frac{1}{10}$ gram, say about 10^{20} ergs; the kinaesthetic sense (e.g. lifting a weight) is coarser still. On the other hand, the eye in rod-vision is sensitive to less than 5 quanta of radiant energy, about 10^{-10} ergs or rather less. In detecting energy therefore man's perceptual apparatus is 10^{30} times more sensitive than it is in detecting mass. Had the perception of mass been as delicate as the perception of energy, the identity of the two would have seemed self-evident instead of paradoxical. When seeing light we should at the same time have *felt* the pressure or impact of the photons; and mass and energy would from the outset have been regarded as merely two different ways of perceiving the same thing. I mention the point here because Dr. Clarke in his criticisms of parapsychology has revived the old arguments based on the *separate* conservation of matter and of energy (see above p. 77, footnote 22). But in any case modern physicists continually emphasize that '*any* conservation law is experimental in origin, and cannot be asserted further than its experimental verification warrants' (D. H. Wilkinson, *ap.* B. J. Blin-Stoyle *et al.*, *Turning Points in Modern Physics*, Amsterdam: North-Holland Pub. Co., 1960).

Special conditions

1. *Subliminal sensory processes.* Before leaving the problems of normal sensory experience, there are three incidental points, germane to our present subject, which call for some discussion. The 'thresholds' for the lower limits of human sensitivity as given in the standard textbooks seem manifestly related to the needs and circumstances of primitive man, and often differ in significant respects from those of other animals. But the figures cited are average thresholds obtained under normal or standard conditions from trained and conscientious subjects with extremely simple stimuli. It is therefore desirable for the parapsychologist to consider the possibilities of heightened sensory or quasi-sensory perception in exceptional individuals or under exceptional conditions. The operative thresholds vary appreciably from one individual to another, and from moment to moment with the same individual. Moreover, they are appreciably reduced when the stimulus is not a simple sound or spot of light but a meaningful pattern; and, as each of us knows, sensitivity may be greatly increased by interest, by practice, and above all, by attention.[26]

It is, however, not sufficiently recognized that occasionally the withdrawal of attention, as well as the concentration of attention, may lower the threshold for sensation, or rather for the reception of sensory stimuli. Everyone is familiar with the maxim that 'the best way to remember a thing is not to think about it'; similarly, there is now abundant evidence to show that sensory processes, as well as memory, may at times function more efficiently when attention is diverted or diminished, and that in such cases they often function unconsciously. The most striking examples of this

[26] Gilbert Murray, whose detailed account of his own telepathic successes in a series of experiments carried out between 1910 and 1920 (*Proceedings, S.P.R.*, Vol. XXIX, pp. 46 ff., Vol. XXXIX, pp. 212 ff., Vol. XLIX, pp. 155 ff.) provides to my mind one of the most convincing pieces of evidence ever published, tells us that 'at one time [he] was inclined to attribute the whole thing to subconscious auditory hyperaesthesia'. McDougall, who was for long a near neighbour of Murray's on Boar's Hill at Oxford, applied the stock laboratory tests, and declared any such explanation was out of the question. Murray himself reports that 'it does not feel like cognition; it is more like the original sense of the word "sympathy"—the sharing of feeling, or "cosensitivity"; . . . it was not so much a piece of information that was transferred, but rather a feeling or emotion. . . . I never had any success in guessing mere cards or numbers.'

type of hyperaesthesia are to be seen in the trance-like states induced by hypnosis. And remarkable results have recently been recorded by experimentalists who have investigated unconscious subliminal perception in normal persons, chiefly with a view to exploiting such processes for purposes of advertisement. It may perhaps be doubted how far the results reported can be taken at their face value; and even if they could, they would scarcely be large enough to alter the general conclusions drawn above, though such possibilities should certainly be borne in mind when planning experiments on so-called extra-sensory perception.[27]

2. *Underdeveloped or atrophied capacities.* If we accept the neo-Darwinian theory of evolution, it would be natural to suppose that from time to time genetic mutations must have produced a rich proliferation of unfamiliar cognitive capacities. If, at the time such variations occurred, they proved useless or detrimental, they would fail to survive on any extensive scale. If, on the other hand, they were helpful under primitive conditions, they might at first have survived and even spread: then, under civilized conditions, when their utility diminished, they would doubtless have atrophied or become suppressed. Many of the animal instincts inherited by prehistoric man have apparently suffered some such fate. And a good deal of evidence has been reported, which, taken at its face value, would suggest that many animals and birds, and certain primitive races, often employ extra-sensory methods of perception.[28] Since civilized people have discovered more

---

[27] The best-attested instances seem to be the surprising skill of the blind in recognizing patterns by touch. From time to time reports are published of exceptional persons who are alleged to possess the power of discriminating colour by touch, or of hearing the minute laryngeal movements made by people when thinking 'silently'; some 'sensitives' profess to obtain detailed information about unknown and absent persons by feeling objects which those persons have possessed or handled; for detailed studies, see E. Osty, *Supernormal Faculties in Man* (London: Methuen, 1923), and the more recent investigations reported by Dr. Hettinger (*The Ultra-Perceptive Faculty*, London: Riders, 1941—a research which gained the award of a Ph.D. from the University of London). With both investigators, however, the experimental techniques employed are open to serious criticism; see D. Parsons, *Journal S.P.R.*, Vol. XLVIII, 1948, pp. 344–52; C. Scott, *ibid.*, Vol. LXIX, 1949, pp. 16–50; and for more favourable comments, C. D. Broad, *ibid.*, Vol. XLIII, 1935, p. 437.

[28] Galton, for example, concluded from observations made during his travels and his later inquiries into visual imagery that these anomalous

effective means of obtaining and communicating information, these side-lines, it is argued, have naturally gone out of use, or survive only in vestigial form. Such suggestions are admittedly little more than speculative conjectures, but they certainly raise problems for more intensive research.[29]

3. *Physical transmission from brain to brain.* In man, when a pattern of sensory stimulation, resulting from the excitation of the peripheral sense-organs, eventually reaches a sensory area in the brain, the *physical* changes in the brain have still to be translated into terms of *mental* changes in the individual's consciousness: only then can we speak of 'observation' or 'perception'. Moreover, it is commonly assumed that the changes in any one individual's consciousness are directly and uniquely correlated with physical changes in that individual's brain. The nature of the correlation remains a mystery; it is utterly unlike any other correlation found in nature. Nevertheless, a long series of experimental studies, carried out during the last hundred years or so, have thrown some gleams of light on the necessary conditions.

Aristotle, partly as a result of observations on living brains, held

---

powers are 'hereditary in certain families, as among the second-sight seers of Scotland, and in certain races, as that of the gypsies', and that they are liable to be 'smothered' by civilized conditions, but may at times be revived by a return to favourable conditions, e.g. 'a life of solitude' ('Visions of Sane Persons', *Roy. Inst. Proc.*, 1881; cf. *Memories of my Life* (London: Methuen, 1908), pp. 273 ff.).

[29] An early and seemingly successful set of experiments on ESP in animals were those carried out by Bekhterev in Leningrad. He claimed to have demonstrated telepathic abilities in a circus dog who obeyed silent commands emanating from his trainer in another room. Pratt subsequently carried out laboratory experiments on similar lines, with apparent success. Still more ingenious investigations have been undertaken to discover whether the 'homing instinct' in pigeons may not be due to extra-sensory capacities; so far, however, the birds have failed to disclose their secret. Even if (as so often happens) such researches throw no light on paranormal powers, they may still succeed in revealing other unsuspected types of sensory perception: many insects, it appears, find their way by polarized sunlight, bats by supersonic echoes; bees make a bee-line to distant sources of pollen and nectar after watching the coded clues danced out by the scout who made the discovery. Sir Alister Hardy, in his fascinating volume, *The Living Stream* (London: Collins, 1965, Lecture IX) has marshalled persuasive evidence and arguments to support the idea that quasi-telepathic processes may have played an important part in biological evolution. See also Chapter VI, pp. 143-164.

that the brain itself was insensible. Even in the nineteenth century it was still widely believed that the brain could only be stimulated from outside, by actual nervous impulses or 'energy' reaching it along the ordinary nerve-paths. However, in 1870 a Prussian army surgeon, named Fritsch, while ministering to wounded soldiers on the stricken field of Sedan, applied an electric current from a portable battery directly to the exposed brain of some of the casualties, and found that this artificial mode of excitation was capable of producing muscular twitches, varying according to the point of application. Of the innumerable experiments since carried out, some of the most remarkable are those of Wilder Penfield. He succeeded in stimulating the bare cortex of conscious patients, and reported that this 'would often elicit quite complex optical and auditory experiences'—visions of landscapes, the sounds of a concert, snatches of conversation, and the like.

Accordingly, still keeping to the ordinary space-time framework of classical physics, we may reasonably infer that, in general, our sensory experiences synchronize (to within a fraction of a second) with corresponding changes in the grey matter of the brain. But we cannot therefore conclude that the 'seat of consciousness' (in the sense of cognitive *activity*) is spatially localized within the percipient's brain. The cognitive *contents* may at times have an apparent location, though they are never located within the brain itself. The pain from a prick on my finger I feel in my finger. The visual sensation, due to a distant lamp that I can see from my bedroom window, I locate on the top of the hill beyond the garden. The light from the lamp is transmitted across space to my retina. But there is no spatial transmission or literal 'projection' of a visual sensation 'generated by the brain' out to the top of the hill. The epistemological relation of perception is not itself a spatial relation, and indeed it is almost meaningless to talk of the cognitive *process* itself as being localized. In a sense it seems to involve 'action at a distance', since I myself am certainly not on the top of the hill. But here the use of spatial language seems wholly inappropriate.

Now, if my consciousness is not spatially located within my skull, then there is no *a priori* reason why it should not, in certain exceptional circumstances, influence or be influenced by processes in the brain of someone else. In such cases B's brain might act as a detector for changes occurring in A's mind, and vice versa. To give the speculation a concrete shape, suppose A and B are identical twins: then the parts or processes of their respective brains which have not been affected by post-natal conditioning, might be so similar that a conscious or semi-conscious state experienced by one might also be simultaneously experienced by the other. There is a good deal of evidence, mostly anecdotal, to support this suggestion. Galton, for example, in his pioneer inquiries reports a case in which, when one twin had toothache, the other experienced toothache in the same tooth. In another case, 'both twins had the same dream at the same hour, 3 a.m.', although their homes were separated by over six miles. In a third case 'one twin, A, who was in Scotland, bought a set of champagne glasses as a surprise for his brother B, while at the same time B, in England, bought a set of precisely the same pattern as a surprise for A'.[30] It is, I suppose, conceivable that in certain respects the brain-structures of persons who have lived in close contact for considerable periods, whether as relatives, spouses, or friends, might also acquire sufficient similarity to function in much the same way. This would then account for the frequency with which such intimate relations between agents and percipients are reported in the various censuses of spontaneous telepathy.[31]

About the time that Fritsch was carrying out his experiments,

[30] *Inquiries into Human Faculty* (Everyman Edition), pp. 162 ff. Later investigators have collected numerous instances of a similar type. When following up the after-histories of monozygotic twins, Miss Conway and I incidentally encountered several stories to much the same effect: but we were not particularly successful in the few experiments we carried out; however, our inquiries were really concerned with very different problems. The remark of one such twin, a lecturer in physics, is worth recording: 'Just as you and your sister exchange letters so as to keep yourselves together psychically though separated physically, so George and I keep tossing to and fro "virtual bits" of information, which binds us together like Yukawa's "exchange forces".' The word 'bit' is here used in its technical sense (a 'binary unit' of information); and the comparison alludes to the way in which nucleons are said to be bound together by the exchange of 'virtual mesons'—a process in which there is a brief violation of the conservation of energy within the limits of the uncertainty principle ('virtual' to a physicist implies a hypothetical process, which from its nature cannot be 'detected' in the classical sense: here the comment was suggested by our failure to obtain confirmation by the usual card-experiments). The comparison was intended humorously; yet at the same time suggests one of many possible hypotheses.

[31] Cf. Celia Green, 'Analysis of Spontaneous Cases', *Proceedings, S.P.R.*, Vol. LIII, 1960, fig. 22, p. 135.

Caton, an English physician, succeeded in demonstrating that the living brain, like other active tissues, generates small electrical voltages. In 1929 a German psychiatrist, Professor Berger of Jena, ingeniously adapted the recording apparatus used to study the electrical activity of the heart, and introduced an amplifier. By this means he succeeded in recording what at first were known as 'Berger rhythms'. They consisted in rhythmical changes in electric potential of about 50 microvolts, with a frequency of about 8 to 13 cycles per second, and are now commonly known as 'alpha rhythms'. Other rhythms varying in strength and frequency were discovered shortly afterwards.[32] It therefore seemed natural to inquire whether the human brain might not be capable of acting both as an electrical transmitter and as an electrical receiver. A particular thought in A's brain might, it was supposed, be accompanied by electrical activity in a certain group of his brain-cells; and a pattern of radiations or 'brain-waves' might then be emitted by A's brain, and (in virtue of something analogous to 'resonance' in the case of atmospheric vibrations) might evoke a similar activity in the corresponding group of cells in B's brain, provided that B's brain was, so to speak, attuned to A's. Now it is interesting to note that 'the resemblance between the "alpha rhythms" of uniovular twins is as close as that of their fingerprints, and the resemblance between the *unstimulated* rhythms persists through the years'.[33] 'Alpha rhythms' have their main source in the occipital areas of the brain cortex, and are suppressed not only by visual stimuli but also by anything which arouses the subject's visualization; so that much the same brain-areas are apparently affected, and much the same brain-processes produced, by stimulation which arises either from a peripheral sense-organ or from within the brain itself—in this case, by any kind of mental activity that involves the formation of visual pictures. Indeed, vivid visualizers can, as a rule, be recognized by the fact that their

[32] 'Über das Elektrenkephalogram des Menschen', *Arch. f. Psychiatr.*, Vol. LXXXVII, 1929, pp. 529 ff.
[33] W. Grey Walter, *The Living Brain* (London: Duckworth, 1953), p. 151. He adds that 'differences soon begin to appear in the details of their responses to stimulation': these 'imposed patterns' are apparently the result of conditioning, and in some degree may be used 'to measure acquired differences of personality'. Thus 'no two people, not even identical twins, have *quite* identical alpha patterns'.

'brain-waves' exhibit smaller rhythmic fluctuations of the alpha type.

The existence of radio-telephony and its application to public broadcasting made the notion of 'Mental Radio' seem highly plausible to the popular mind; and the theory was championed with much crude and picturesque detail in a book by Upton Sinclair with that title. Parapsychology appears to have followed the same model. Contemporary science, however, tends to think in terms of interactions rather than of mere transmission. And I am inclined to suggest that parapsychologists would do well to adopt this change of tactics, and frame their hypotheses on the assumption that the processes involved are essentially reciprocal, although in many cases, if not in most, one partner—sometimes the agent, sometimes the percipient—may play the predominant role. It was, I fancy, something very like this principle of reciprocity which Jung had in mind when he put forward his theory of 'non-causal synchronicity'.[34]

In an earlier paper I suggested that what Jung calls a 'scheme of experience' might be tentatively interpreted in terms of the 'interaction' of two psychophysical 'fields' (i.e. of two minds conceived not as confined within the corresponding skulls but spread out into space) which were 'wholly or partly isomorphic'. Whether or not such an interaction occurred in any given case would depend not merely on the states of the persons involved but also on the 'situation' obtaining in the wider 'topological field' (in Lewin's sense) in which the two 'personal fields' were temporarily placed.[35] If we think of the isomorphism in physical terms as primarily dependent on similar brain-structure, then we might expect the ideal conditions to be found in the case of identical twins. Indeed, among the few experiments of my own which

[34] C. G. Jung *Naturerklärung und Psyche* (Zürich: Rascher, 1952). 'The connection of events,' he writes, 'may in certain circumstances be other than causal; hence a second principle of connection is required.' He suggests that the process occurs when the total situation is 'meaningful'. In such cases the process involves no real 'transmission', but rather what ancient writers referred to as 'sympatheia'.

[35] This formulation was suggested, partly by experiments of my own, and partly by the details of the telepathic experiments carried out by Soal and Bowden on two Welsh brothers (C. Burt, 'Experiments on Telepathy in Children', *Brit. J. statist. Psychol.*, Vol. XII, 1959, pp. 87 ff.; S. G. Soal and H. T. Bowden, *The Mind Readers*, London: Faber, 1959).

appeared to indicate a significant measure of success, the most successful of all were obtained with monozygotic twins.[36] Berger himself considered the possibility in all seriousness, but eventually rejected it.[37] An Italian neurologist, M. F. Cazzamali, however, claims to have recorded 'electro-magnetic waves', capable of conveying information direct from one brain to another; but both his methods and his inferences are open to criticism.[38]

The most systematic attempt to develop and test a plausible physical hypothesis of this type is that of Dr. Vasiliev, Professor of Psychology at Leningrad. Bekhterev, his teacher, who with Pavlov was the leading Russian exponent of 'human reflexology', believed that 'the emanation of electro-magnetic waves would provide the best working hypothesis to account for telepathic phenomena'; and it was this tentative conjecture that Vasiliev set out to investigate. His method was to enclose his most sensitive subjects in a metal 'Faraday cage' in order to screen off all such waves. He found only a small diminution in their telepathic achievements, too slight to be statistically significant.

[36] See the paper just quoted, p. 88. F. W. H. Myers had already remarked that 'one clue to the causes directing and determining telepathic communication lies in what seems the exceptional frequency between *twins*—the closest of all relations', and elsewhere notes that 'Gurney in his analyses of relationships in telepathic cases made the interesting observation that the link of *twinship* seems markedly to facilitate this kind of relationship' (*Human Personality*, I, Sect. 427, p. 144 and Sect. 653, p. 272 and refs.; his italics).

[37] *Nervenkr.*, Vol. LXXXVII, 1929, pp. 527 ff. Berger's views on the relations between the electrical phenomena and thought-transference are described in a monograph entitled *Psyche* (Fischer Verlag, 1940). Here he briefly relates some of his own telepathic experiences, and describes a series of experiments carried out with various subjects, often under hypnosis. He states that he himself was 'a poor transmitter but a good receiver'. Since the most conspicuous of the rhythms, the 'alpha' waves, are *inhibited* by visual and other mental activities, he argues that they cannot themselves act as carriers; but he supposes that the electrical energy might be 'transformed into psychic energy', and that this is radiated in the form of waves. In recent research on encephalography one of the most fruitful techniques has been the use of flickering lights which (under certain circumstances) may alter the rate of the cortical rhythms so that they keep in step with the flicker (see W. Grey Walter, *op. cit.*, Ch. IV). This would seem to be a type of brain-wave, which could readily be identified if transmitted from one brain to another, and might well provide a further check.

[38] *Reports to the International Congress*, Paris, 1927, and Oslo, 1935. Berger observes that he failed to detect any radiations of the type described by Cazzamali.

In later experiments he varied the distance between agent and percipient from 25 metres to over 1,700 kilometres (Leningrad to Sebastopol). There was no appreciable weakening of the 'signals', and no increased delay in the response. It therefore appeared that, at any rate in the form proposed, the hypothesis of a 'physical field' was highly improbable.[39]

The difficulties attending a theory of electro-magnetic transmission are obvious. (i) Vasiliev, like most earlier writers, assumes that the 'electromagnetic radiations from the brain', like light from an electric lamp, are emitted in all directions. In that case their intensity would diminish in accordance with the law of inverse squares. One would therefore expect the rapidity and strength of the recipient's response to diminish appreciably as the distance is increased. Nothing of this sort is found in the *ad hoc* experiments devised to test this point. In my view, however, too much weight is laid commonly on this type of argument. If the receiver's brain includes an amplifier and detector, it might also contain an automatic volume-control to compensate for loss of signal strength. Or the receiver's brain might include a mechanism which, like the laser, might produce amplification by stimulated emission of the relevant radiation. The radiation might be concentrated and directed in an almost linear fashion towards the recipient, and only when the direction was accurate would information be received. (ii) Far more to the point, I fancy, is the fact that the only known 'electro-magnetic radiations from the brain' are those recorded in encephalography. These have rhythms that

[39] L. L. Vasiliev, *Experiments in Mental Suggestion* (London: Inst. Mental Images, 1963). His cage, it may be noted, would not have impeded the transmission either of extremely short waves (hard X-rays and gamma rays) or of radiations having a wave-length of over 1 kilometre. Other Russian physicists have carried out researches along similar lines which merit close attention. The renewal of Russian work on parapsychology was apparently due to a journalistic report (since contradicted) describing successful experiments on the United States submarine, *Nautilus*. The interest of military authorities in the possibilities of extra-sensory perception is of very ancient standing. When Benhadad the Syrian complained that his plans seemed always to be known in advance to his enemies, he was told, 'Elisha, the prophet that is in Israel, telleth the king the words thou speakest in thy bedchamber,' and accordingly took steps to capture the telepathic expert (2 Kings VI. 8-13). It was doubtless with the military possibilities in view that Vasiliev decided to study the transmission not of thoughts but of commands.

are extremely slow, ranging from 5 to 30 cycles per second. Thus even the fastest have a slower frequency than the lowest note on the piano. This implies wavelengths of 10,000 to 20,000 miles. It is extremely difficult to conceive that radiations of this type could serve as carriers of information. (iii) The changes in electrical potential are of the order of 5 to 50 microvolts. Hence the power involved must be in the region of 10^{-20} watt. Such changes could not be detected by the most delicate physical apparatus at a distance of more than an inch or two from the skull. Propagation to the distances reported for telepathy would require far greater power—10 watts at the very least. (iv) Finally, the transmission of information would require some code-like pattern to be imposed on the electro-magnetic radiations, i.e. some kind of 'modulation'; and the receiving brain would need, not merely to detect and amplify, but also to decode, the message conveyed. In the actual structure of the brain there is nothing whatever to suggest a mechanism of the sort that would be necessary for such purposes. And if there were, one would expect that certain brains at any rate would be affected by other forms of electro-magnetic signalling.

The foregoing objections, together with the experiments I have described, would seem completely to rule out the ever-popular theory of 'brain radio'. They do not, however, wholly exclude all other physical modes of transmission. The brain is by far the most complicated material structure in the physical world; and the pyscho physical processes that take place in it are of a kind found nowhere else. It therefore remains conceivable that processes so unique might give rise to the emission of some form of energy hitherto unknown—a form presumably to be found only in structures associated with mental (i.e. potentially conscious) processes. In that case many psychologists would doubtless claim that the energy itself was mental; but the question is mainly verbal. The advantage of regarding the communication as essentially physical would be that our hypotheses could then be framed on an analogy with processes already intensively studied, and might consequently lead to corollaries that could be tested by well-tried experimental procedures. However, once the relatively simple theory of electro-magnetic radiation has been discarded, any other physical hypothesis of the same general type would need to be invented *ad hoc*, and could therefore claim no more than the low *a priori* probability of all such specially improvised conceptions.

I am, however, tempted to suggest that the notion of paranormal experiences as consisting typically of a one-way transfer of information is itself a mistaken starting-point. The whole process seems far more complex. Even in ordinary perception, as Stout and other apperceptionists have never wearied of insisting, the perceiver contributes as much as the percipiendum; and it takes a couple to render a communication effective—an appropriate recipient as well as an efficient communicator, and the situation obtaining must also be favourable.

Kant, we recall, distinguished two ways in which one phenomenon might determine another: the first involves 'the asymmetrical relation of cause and effect', in which the two events are separated by an interval of time; the second involves 'the symmetrical relation of reciprocity', in which the events are simultaneous.[40] Nineteenth-century science, particularly in its more

[40] I. Kant, *Kritik der reinen Vernunft*, 'Transcendentale Analytik, Dritte Analogie; Grundsatz des Zugleichseins, nach dem Gesetz der Wechselwirkung oder Gemeinschaft.' Kant's 'transcendental analytic' was essentially an attempt to state and examine the metaphysical presuppositions underlying the physical science of his day, which was derived chiefly from Galileo, Huyghens, and Newton. The reason why he assigns so prominent a place to the principle of 'reciprocity' is that Newton had added the third law of motion to the two which he took over from Galileo, and thus in his theory of gravitation erected reciprocity into a cardinal principle in physics ('as shown by the phenomenon of the tides, not only does the earth attract the moon, but the moon attracts the earth'). It is odd to find commentators, even as late as 1939 declaring that 'for us this part of Kant's metaphysics has only a historical interest' (R. G. Collingwood, *Essay on Metaphysics*, London: Oxford University Press, 1940, p. 270). I have already alluded to the nuclear physicist's hypothesis of 'exchange forces' (which are quite unlike the causal forces of classical physics): they make their appearance in a wide variety of contexts—chemical binding, the cohesion of solids, the properties of crystals, and a number of other so-called 'co-operative' phenomena. But the principle of reciprocity is invoked to exclude as well as to bind. We are all familiar with the scholastic dictum, that, just as two *mental* substances (or, as we should now say, two personal consciousnesses) cannot share the same body, so two *material* bodies cannot occupy the same space. In much the same way there is a similar but somewhat surprising 'exclusion principle' in subatomic physics, which states that two electrons cannot share the same state of motion (or in more technical terms possess the same four quantum numbers). These mutual relations, as Margenau puts it, seem to 'regulate the *social* behaviour' of elementary particles, and, unlike the mechanistic theories, 'may provide a

popular and practical forms, laid all its emphasis on the causal type of action. Parapsychology seems to have followed suit. Twentieth-century science tends rather to think in terms of interactions.

Conclusion

It would appear then that, even on the simpler, lower plane of direct empirical investigation, where systematic observation or experiment can be used, there is no lack of physical hypotheses to be tested and checked, nor of appropriate laboratory techniques for confirming or refuting them—at any rate (and the proviso is important) so long as we assume that the alleged paranormal phenomena are intrinsically sensory in their nature or content. There can therefore be no cogent grounds for rejecting the notion of extra-sensory perception from this somewhat limited standpoint. Even if, with Professor Hansel, we agree that any form of extra-sensory communication would defy interpretations in terms of the 'laws and concepts' of classical physics (i.e. the physics of mesoscopic phenomena), it is quite easy to construct a variety of plausible explanations in terms of the 'laws and concepts' of subatomic physics.[41]

Nevertheless it seems equally clear that these physical hypotheses could at most only account for the conditions (or some of the conditions) which need to be fulfilled if extra-sensory phenomena are to manifest themselves; they do not explain the process itself. The mere fact that a hypothesis is constructed in accordance with some physical model does not imply that the whole activity is essentially physical. Kelvin, it may be remembered, pointed out that the mathematical formalism devised to describe the flow of fluids could also be used to describe the conditions and laws governing 'the flow of heat, the phenomena of electrical transmission, of magnetism, and even of elasticity'; at the same time he insisted that no specific physical explanation would necessarily

---

[41] For details see *Brit. J. statist. Psychol.*, Vol. XII, 1959, p. 86 and references.

key to various biological problems' (including psychological): they furnish, so to speak, 'a way of understanding why entities show, in their togetherness, laws of behaviour quite different from the laws which govern them in isolation' (*The Nature of Physical Reality*, pp. 442 ff.).

follow from the successful confirmation of such analogies: Fourier did not use them to maintain that heat was a material fluid; and Maxwell emphasized that, 'in treating lines of force as lines of flow in an incompressible fluid', he did not mean to suggest that 'electricity itself was just another fluid'. It is a basic assumption of communication theory that information can be treated very much like a physical quantity, such as mass or energy, which can be conveyed from a source to a receiver by means of a channel with a specifiable capacity. Suppose then that we found that, in those aspects which can be empirically tested, the flow of information from the brain of one telepathic twin to the brain of the other resembles 'the flow of a fluid of varying viscosity along tubes of varying calibre', it would not follow that what was actually conveyed was in fact some *physical* quantity—e.g. 'an outward radiation from the brain of extremely short and penetrative waves' (Crookes' suggestion), or 'some unknown type of energy into which the energy of the brain can be converted' (Berger's suggestion), or again 'some psychokinetic influence capable of triggering off any neurones in the percipient's brain that are in a critical state of sensitivity' (Eccles' suggestion).[42]

Above all, there would still remain what I have called the

[42] J. C. Eccles, *Nature*, Vol. CLXXV, 1951, pp. 516 ff. Dr. Taylor thinks that, in my references to Eccles' theory (*loc. cit.*, p. 75 and *passim*), I have 'misunderstood the notion of psychokinesis'. It is true that Eccles describes the 'extraneous influence' as exerted by what is loosely called 'mind' or 'will'; nevertheless he insists that it must have a spatial or spatio-temporal pattern, but I doubt if he is thinking of it as 'quasi-physical'. Let me add that if what is transmitted were in actual fact physical (in the ordinary sense), it should not be beyond the wit of the laboratory expert to devise some means of intercepting it *en route*. Hitherto, however, apart from the 'electromagnetic theory', nearly all such physical hypotheses have been expressed in a form too vague to be susceptible of experimental testing.

Since the above was written an extremely attractive suggestion has been put forward by Mr. H. A. C. Dobbs. This is based on the ingenious idea that telepathic communication might be effected by what he calls 'psitrons'—particles of 'imaginary' rest-mass travelling across the intervening space with a velocity exceeding that of light ('imaginary' in the mathematical sense). If captured by cortical neurones in a sufficiently susceptible state, they might then trigger off a chain reaction of neurone-discharge. The whole fascinating theory is formulated with sufficient precision to suggest a clear-cut experimental programme for testing both the theory as a whole and its various tentative details ('Time and ESP', *Proceedings S.P.R.*, Vol. LIV, 1965, pp. 339 ff.), and see Ch. X in this book.

epistemological problem. A suitable receiver, suitably attuned, can accept and record a message transmitted by wireless; but the physical receiver never becomes *aware* of the message. The human brain (or 'mind') not only receives and records the message but in the most striking cases becomes conscious of its nature. There is not merely extra-sensory *communication* but also extra-sensory *cognition*. It should, I venture to suggest, be the special task of the psychologist to make an intensive study of this final link in the telepathic chain.

Most investigators adopt a far too naïve view. Their tacit assumption would seem to be somewhat as follows. When the agent looks at a specific object (the star, say, on a Zener card), a patterned excitation occurs in his visual cortex, which then projects a distinctive change across the intervening space. On reaching the percipient this complex signal arouses a similar excitation in a corresponding part of the percipient's visual cortex, provided only that its strength on arrival exceeds the momentary threshold-value imposed by the ever-varying condition of the relevant nerve-cells. The result is a sensory percept or sensory image of the same type as that experienced by the agent. Now, as I have argued in an earlier paper, judging by the introspective comments of my own subjects and the incidental observations recorded by other investigators, it would seem to be quite exceptional, even in spontaneous cases, for the percipient to perceive or imagine a sensory content *which actually resembles* that perceived or experienced by the agent. The whole cognitive process appears to be far more subtle and indirect, and to a large extent unconscious. 'In all such cases,' I contended, 'it seems more probable that what we have to deal with is, not percept-transference or even thought-transference, but a mode of implicitly guiding or directing thought so that the so-called "percipient" *thinks* of the right object or the appropriate situation at the right moment.' (This, I fancy, is what many writers mean when they say that the transmission comes first to the 'unconscious mind', to the 'subliminal self', or to a kind of 'group subconsciousness'). Then, much as in dreams, the 'latent content' derived from the message may get translated into a 'manifest content'—usually concrete, often visual, occasionally auditory (as in hearing voices), more often symbolic than strictly representative, and sometimes merely emotional. But quite often it may be simply converted into appropriate actions or utterances

—as in card-guessing, where the guesser is usually unaware of receiving any message whatsoever.

Here certainly, so at least it would appear, the brain functions, not by directly generating the contents of consciousness but by selecting them: 'in transmitting information to the conscious mind' (so I argued) 'it commonly permits only what is absolutely essential to percolate through'. At its lower, semi-conscious or unconscious levels it continually exercises a kind of censorship, shutting out what has no obvious bearing on the practical issues of the moment. It is this process of censorship, I believe, that accounts for the perceptual decline of paranormal powers and of paranormal phenomena to which Professor Hansel and other sceptics so scathingly refer. 'There has never since been a medium' [we are told] 'who could achieve the success attributed to Mrs. Piper, nor have the remarkable coincidences observed in the early days of "cross-correspondences" ever been repeated. . . . Neither Rhine nor Pratt have succeeded in living up to the reputation they achieved with their early experiments on card-guessing; and whenever a fastidious critic attempts to verify for himself the results attained with Shackleton or the Jones boys, he is told that their powers have since deteriorated.'[43] There is certainly an element of truth in these and similar remarks. As soon as a man or woman—or, what in my experience is still commoner, an adolescent youth or girl—begins to realize that he is liable to extra-sensory, paranormal, mystical, or any other exceptional types of experience, his mind usually seems to resist or suppress

[43] One could of course retort that each pronouncement of failure is commonly followed by another report of remarkable success. Soal's prolonged failure was followed by unexpected successes. Miss Cummins' recent 'study in automatic writing' (*Swan on a Black Sea*, London: Routledge, 1965) is every whit as impressive as any of the earlier studies of equal length. Mr. Milan Ryzl's subject is still apparently yielding high scores with a number of independent investigators from different countries. Dr. Taylor, however, replies that 'even so, in comparison with the numerous investigators now engaged on psychical research in all parts of the world, the positive results remain not merely unconvincing but surprisingly few and far between'. Let him note what Bridgeman says of research in more orthodox fields of science: 'There is an aspect often lost sight of—the small proportion of successful discoveries compared with the number of investigators' (*The Logic of Modern Physics*, London: Macmillan, 1946, p. 209); and let him recall Faraday's remark, that he was 'satisfied if only one-tenth of a percentage of his experiments turned out to be successful'.

them, partly for practical reasons, partly for social. 'The brain blocks them off; only in certain more or less abnormal states—states of dissociation or the like—do they at times still force their way out, and become active. The same holds true of groups and communities: the material and utilitarian outlook of civilized life almost inevitably forbids any diversion of attention to the remoter planes of cosmic existence.'[44]

III LIMITATIONS OF CONCEPTUAL PROCESSES

Arguments like the foregoing, which are based chiefly on observational evidence, are essentially empirical. Consequently, for those who reject the parapsychologist's conclusions on *a priori* grounds they are unlikely to carry much weight. The principal objections urged by such writers arise, as we have seen, from more general considerations. Briefly their main contention boils down to this: the very notion of a paranormal process is incompatible with the clear and comprehensive concept of the universe which, after three centuries of scientific inquiry and mathematical analysis, the human intellect has at last achieved.

And here, strangely enough, it is the psychologists themselves

[44] I have dealt with this process more fully in a paper on 'The Structure of the Mind' (*Brit. J. statist. Psychol.*, Vol. XIV, 1961, pp. 168 ff., from which the last quotation is taken). As I observed in my discussion of the Welsh boys studied by Soal, those whose memories are old enough, and who visited the less frequented corners of Snowdonia and possessed some ability to gossip in the native tongue, can recall how, half a century ago, local opinion took extra-sensory phenomena as a matter of course. Some years ago I had occasion to return to the same villages, and found that these beliefs had almost entirely died out. A hint of this repressive tendency finds an utterance in such familiar comments as 'it just wouldn't do if we could read other people's thoughts' or 'if we could see into the future'. And, as every practical psychologist is only too well aware, this kind of repression crops up in other fields besides those of parapsychology. Again and again, when I have lectured on hypnotism and found a student who could serve as an admirable subject in demonstrations, his susceptibility would vanish as soon as he heard from his fellow-students what an excellent performance he gave. Businessmen who display amazing shrewdness in summing up the capacities or the motives of comparative strangers may be completely blind to the characteristics or qualities of their wives or children. Gifted youngsters from the working classes, who seem full of intellectual promise, often seem to lose it all as soon as they go out to work; and one is tempted to echo Brutus's reflection:

> *What a blunt fellow this is grown to be!*
> *He was quick mettle when he went to school.*

who are most emphatic. Mr. Hammond, for example, declares that 'parapsychology, like the older brands of dualistic psychology which we have long since discarded, is flagrantly at variance with what we know of the Unity and Uniformity of Nature'. Dr. Taylor puts his faith in 'the principle of simplicity', and says 'what logic requires is that we should use Occam's razor, not only to lop off the *ens* called "mind", but to eliminate any pseudo-scientific concept which is not already part of the current concepts of natural-scientific discourse'. Professor Hansel, as we have seen, maintains that 'all the illusory notions elaborated by psychical researchers can be dismissed out of hand as impossible', because they 'violate the basic requirements and principles of natural science'. And finally, Mrs. Knight sums up these various objections by tersely observing that 'the facts alleged by parapsychologists just cannot be organized within the accepted scientific framework'.[45] With these writers therefore—and, I think we may say, for the vast majority of sceptics, lay as well as academic—the chief reason for rejecting the conclusions of parapsychology is not that they fail to stand up to observational or experimental tests, but rather that they are inconsistent with some comprehensive conceptual scheme—a scheme which they treat as the final court of appeal.

In considering this more radical line of attack the first point to make is that the conceptual activities of the human mind are as imperfect and as limited as its powers of sensory perception, and the conceptual procedures of the scientist are the most limited of all. The world of our *actual* experience consists, not of general concepts but of individual entities and individual events, each utterly unique. This experience is for all of us (including every scientist) what each of us describes as '*My* experience'; it is made up of what happens to '*Me*', as '*I*' sit on a particular chair in a particular room or go about my daily business. All reality is concrete; and the story of the universe in which you and I live, like the story of my life or your life, is a piece of individual history, not an abstract system of mathematical equations or universal laws.

And yet, if he is to cope with such a world, the scientist is

[45] *New Scientist*, Vol. 1959, pp. 457 ff.; *Brit. J. statist. Psychol.*, Vol. XIII, 1960, pp. 175 ff., Vol. XVII, 1964, pp. 73 ff.; *Science News*, 1950, p. 9; J. G. Taylor, *The Behavioral Basis of Perception* (New Haven and London: Yale University Press, 1962).

obliged to start by classifying the individual entities into generalized sets, categories, or types, and by reducing the individual events to broad recurrent processes conforming to more or less general rules; and in doing so, as Ward observed, he 'tends to forget that all experience is essentially historical'.[46] In the physical sciences, the inadequacies of this reduction of the concrete to the abstract, of the particular to the general, do not as a rule greatly matter. But in psychology and especially in parapsychology the scientist's habit of conceptualizing and generalizing may lead him seriously astray. The most important of all paranormal phenomena owe their significance to the very fact that they were specific events which happened to particular individuals—like the visions of Isaiah, St. Paul, or Pascal, or the voice which St. Augustine heard in the garden, or the ghost of Marley seen by Scrooge on Christmas Eve.

Here, however, our business is with the scientific study of such phenomena. And for this purpose, I agree, we must, so far as we are able, comply with the 'basic requirements and principles of natural science', or, as I should prefer to say, its basic methods, all the time bearing in mind the limitations of such methods. Of these the one on which the greatest and most frequent stress is laid is the principle of simplicity or economy. Mr. Gregory formulates it succinctly when he says that 'a pure or ideal science should form a coherent deductive system which will enable the maximum number of observable facts to be inferred from a minimum number of unproved assumptions'.[47]

Now the requirement that the 'unproved assumptions' should be pared down to the barest minimum provides a sound methodological rule when we are constructing a positive *proof*, because by adding to their number we manifestly increase the vulnerability of the inferences deduced from them. But to turn such a principle into a major factual premiss, and use it as a method of *disproof*, is

[46] J. Ward, *Naturalism and Agnosticism* (London: A. & C. Black, 1899), p. 571.
[47] W. L. Gregory, 'Scientific Method in Psychology', *Brit. J. statist. Psychol.*, Vol. XI, 1958, pp. 97 ff., and R. B. Braithwaite, *Scientific Explanation* (London: Cambridge University Press, 1953). See also G. Schlesinger's chapter on 'The Principle of Simplicity' (*Method in Physical Sciences*, London: Routledge, 1963), which cogently disposes of 'the thesis of Nature's simplicity', as held by Galileo, Kepler, and revived by R. O. Kapp (*Brit. J. Phil. Sci.*, Vol. VIII, 1958, pp. 285 ff.).

wholly indefensible. It is, moreover, of little value when we are seeking to *describe* phenomena in all their kaleidoscopic detail, particularly phenomena in a domain that has hitherto been but little explored; and even the physical scientist is quite ready to abandon it when he is seeking to *understand* things.

'Pure science' is, and always has been, largely an outgrowth of 'applied science,' i.e. of technology. The problems it sets out to solve, and the range of the 'observable facts' it consequently embraces are nearly always prescribed by current practical interests and current practical needs. Hence its predominant aim has been, not so much to describe or to understand things, but rather, as the behaviourist is so fond of insisting, to 'predict and control' them. Its method therefore is not to hold a mirror up to nature but to devise the simplest possible working model—in one word, a calculus, preferably a mathematical calculus since that is at once impartial and precise.

For this purpose it proceeds, as before, by selection and rejection. What the scientist observes is, by the very fact of observation, inevitably related to an observing or cognizing 'subject'. But both 'subject' and the cognizing relation are treated as irrelevant, and expunged from the ensuing description; and in this way science professes to become purely 'objective'. Furthermore, all those aspects or contents of experience, which seem unnecessary or inconvenient for an abstract calculus, are also cast aside at the outset—all emotional, aesthetic, and moral elements, and what are nowadays termed 'values'. 'Secondary' qualities have likewise proved useless for generalized predictions; those, too, are therefore swept into the scientist's dustbin. Only those variables are retained which lend themselves to quantitative treatment and to mathematical deduction.

Physicalism and behaviourism

Let us now look more closely at the nature of the particular conceptual system which has thus been hammered out—'the accepted scientific framework', to which the critics of parapsychology almost unanimously pin their faith. Science naturally began with the study of man-sized objects and their more obvious observable relations; and in his efforts to predict and control their behaviour the scientist has constantly had to ask—what will happen later on if I proceed to do this now? Hence from the very start the concepts of time and causation acquired a paramount importance. For a working interpretation of these gross 'mesoscopic' phenomena, as I have called them, familiar concepts sufficed—the spatial scheme supplied by vision, the notion of matter defined in terms of weight (or rather mass), and of force conceived in terms of push and pull. Hence the early emergence of the science of mechanics, and its remarkable development during the nineteenth century, which was typically an age of engineering. The steam engine did far more for science than science did for the steam engine. The climax was Lord Kelvin's famous criterion [48] that 'no theory could be deemed worthy of scientific consideration unless it was possible, at least in principle, to make a mechanical model of it'; and this test was applied even to theories of the universe itself. It survives, as we have seen, in Professor Hansel's declaration that the conclusions of parapsychology can be 'dismissed out of hand' because they 'conflict with the laws of mechanics'.

As a result of these early triumphs in the technological field the 'scientific framework' ultimately accepted by the majority of physicists during the closing half of that century was rigidly mechanistic. It was attained by pressing the principle of simplicity to its farthest limit and postulating only a single concept for every traditional category—in short, a thoroughgoing monism. It assumed a single universal container, namely, three-dimensional Euclidean space; and this was supposed to contain only a single type of substance, namely, matter in shape of indivisible, indestructible, and immutable atomic particles, each characterized by a single property, namely, mass, controlled by a single type of cause, namely, mechanical force operating by contact—all to be verified by a single type of observation, namely, measurements from 'pointer-readings,' with the result expressed by a single type of proposition, namely, a differential equation. Action at a distance (e.g. the pull of gravity and magnetism) were to be accounted for by postulating one further entity—an elastic, jelly-like ether, pervading all space, and subject to strain, tension and undulatory waves. Since all events are causally determined, the whole past and future of the universe could in principle be deduced and predicted from a single set of measurements with the exactitude of the Nautical Almanac.

[48] *Nature*, Vol. XXXI, p. 603.

So far as man-sized objects and processes are concerned, this simple methodological assumption furnished surprisingly reliable results, and the dazzling successes thus achieved blinded men's eyes to its obvious limitations. Save for a few obstinate exceptions, which, it was hoped, would soon be explained away, scientific writers of the late Victorian era felt convinced that in its essential outlines the 'framework' of the universe had been accurately marked out; all that remained was to fill in the details. The theory of natural selection, and the rapid progress made by genetics, biochemistry, and molecular biology, have since brought all the phenomena of life within the same materialistic scheme, or so it is commonly believed.[49] If the human brain still presents a tangle of baffling problems, that, we are told, is due to a difference not in kind but only in complexity. The clue to its general working, so the earlier behaviourists assured us, was disclosed with the discovery of the 'reflex arc'—the sensori-motor nerve-circuit. Just as incoming and outgoing telephone-calls can be explained in terms of wave-like currents running along appropriately arranged wires when a dial is turned at the transmitting end and the appropriate connections are joined at the central exchange, so

[49] The most trenchant statement is that of T. H. Huxley: 'The human body itself is a machine, all the operations of which will, sooner or later, be explained on physical principles. . . . All states of consciousness are caused by molecular changes of the brain substance. I believe that we shall arrive at a mechanical equivalent of consciousness, just as we have arrived at the mechanical equivalent of heat' ('On the Physical Basis of Life', *Method and Results*, 1893, pp. 160 ff.). The present-day behaviouristic view, as expounded for example by Dr. McLeish, is precisely the same except for slight changes in terminology.

It should, however, be noted that there is more than one expert biological critic who questions the current doctrine of purely mechanistic theories of life and evolution: see Sir Alister Hardy, *The Living Stream*, and W. H. Thorpe, *Science, Man, and Morals* (London: Methuen, 1966). Both they and L. L. Whyte (*Internal Factors in Evolution*, London, Tavistock Publications, 1965) deny that the mechanical theory of natural selection (the elemination of ill-adapted variations by the external environment, with the survival of the better adapted) is alone sufficient to explain evolution. Internal factors, making for an increasingly elaborate and harmonious organization of the variant elements in the individual organism as a co-ordinated whole, also play an essential part. Hardy's theory represents the organism as able to choose a new and more complex environment to which it is already adapted instead of waiting for blind environmental selection; it is a theory particularly well suited to the explanation of *mental* evolution.

the stimulation and reactions that make up human behaviour might (it was supposed) be explained in terms of somewhat similar waves of electro-chemical activity flowing up, down, and along the nerve-fibres which lead to, across, and from the brain—waves or impulses which are in fact accompanied by observable electrical effects. Some of the connections are built in from the start (the 'inborn reflexes'); others are laid down by 'association', or, as the Pavlovians have taught us to say, by 'conditioning'. Man himself, it would appear, is nothing but a complicated automaton, and so after all offers no exception to the general mechanistic scheme. 'I want,' wrote Professor Mace in his admirably lucid exposition of the behaviourist theory, '*one* world of material things, and *one* science to deal with all the things that this world contains';[50] and both, so he maintains, have now been provided by the conception which regards the brain as a mechanical apparatus for processing information, very like the so-called 'electronic brain'.

The idea of the universe as a gigantic machine, with human beings as minor parts of the clockwork, is so attractively simple that already, before the Victorian era had ended, most scientists felt that they could at last comprehend the universe and its manifold workings. This neat mechanistic theory 'supplied the conceptual background in which', J. B. Watson, the apostle of behaviourism, (as he himself has related) 'was nurtured from early youth'; and owing to its extreme simplicity it rapidly filtered down to the non-scientific members of the public, and furnished an intelligible *Weltanschaung* for what was described as 'emancipated common sense'. In such a universe any unsophisticated

[50] C. A. Mace, 'Physicalism', *Proc. Aristot. Soc.*, Vol. XXXVII, 1937, p. 24. Professor Mace, who writes not as a physicist but as a professor of psychology, has since considerably modified his views; but they express in clear and logical fashion a conception of psychology which still very widely prevails. It need hardly be said that the demand for 'one world of *material* things' can no longer be taken literally. Long before Einstein had deduced his equation asserting the equivalence of matter and energy, matter as such had already been dematerialized; and the crude type of nineteenth-century materialism still postulated by Hansel and several of his fellow-critics was already, as Eddington used to say, 'not only damned but dead'. Mace himself as the title of the paper indicates, uses the simple phrase 'material things' to denote 'observable systems or constructs which can in principle be reduced to terms of concepts and the laws of physics'.

thinker could feel comfortably at home. Most people, however—even scientists and psychologists—have, I suspect, been led to adopt this simple creed more from half-unconscious motives than from any first-hand study of the empirical evidence or of the philosophic postulates on which it was based; and that makes it all the harder to root out. Clearly, so long as they regard this mechanistic metaphysic as the last word on the subject, critics like Professor Hansel and the others whom I have cited, are bound to conclude that the antecedent probabilities are wholly against the assumption of immaterial entities like minds, and *a fortiori* against taking ostensibly paranormal phenomena at their face value.

Here, then, we have the main and commonest reason for rejecting parapsychology. We have now to consider how far this mechanistic theory will stand up to the objections that can be urged against it.

Objections to physicalism

1. *Common sense.* For the plain man the obvious objection to a strictly physical hypothesis is that it flies in the face of all common sense. The homogeneous type of universe that it postulates seems quite incompatible with the heterogenous qualities and processes of everyday experience—the rich variety of colours, sounds, and smells that characterize the human scene, the clear distinction between conscious creatures on the one hand and the inanimate objects of the material world on the other, above all, the ineradicable sense that we ourselves possess the power of choice and will. Even those who in their official moments are uncompromising exponents of physicalism, nevertheless, in their ordinary domestic life (as anyone who knows them can testify) would never dream of treating their wives, their children, or their sweethearts as automatic dummies, or excusing their own faults as the aberrations of a physico-chemical machine. Over the teatable the most hardened behaviourist continues to gossip, *malgré lui*, about his conscious experiences, his feelings, and his intentions like any other mortal.

2. *Logical.* Pressed to its logical conclusion, the contention of the thoroughgoing mechanist is, quite palpably, self-destructive. He maintains that all human actions are determined, not by

conscious reason or voluntary choice, but blindly and mechanically by physical processes within the brain. It follows that his own actions, including his pronouncements on the problems of psychology, must be blindly and mechanically determined. On his own showing, therefore, he cannot help saying what he does say. Why, then, should we take the smallest notice of what he asserts? However, mere *tu quoque* rejoinders are a feeble method of defence; and it is incumbent on the supporters of parapsychology to examine what precisely is the factual evidence on which their critics claim to rely.

3. *Physical.* The empirical arguments which they commonly bring forward follow two main lines: first, we are told, 'the theories of the parapsychologist arbitrarily and needlessly multiply' what Dr. Mumford calls 'the basic *contents* of Nature'; secondly, and just as arbitrarily, they 'violate the established *laws* of Nature'. And by 'Nature' is evidently meant 'nature' as interpreted by natural science in its present stage.

(i) It is obvious that in a purely mechanistic universe anything like a mind would be a most incongruous intruder. However, such concepts as mind, and particularly disembodied minds, and such notions as purely mental processes, particularly processes like telepathy or clairvoyance, are (so the behaviourists assure us) merely attenuated relics of the old Cartesian dichotomy; all such 'dualistic theories', we are told, 'gratuitously disrupt the unity of the universe by postulating two types of entity and two types of activity where one should suffice'. But to argue that the unity of the universe implies that its contents must be homogeneous is like arguing that a picture cannot form a harmonious whole unless it is painted in monochrome. Yet why, the critic may still protest, need the present-day psychologist try to introduce such elusive notions as minds and mental processes into a world which, *prima facie* at any rate, consists of nothing but physical objects and purely physical processes? The answer is simple. A purely physical process is an unconscious process; and, whether or not they spoil the unity of the universe, conscious processes incontestably exist. Even in the physical world, despite more than a hundred years of ingenious theorizing, the attempt to achieve a unified interpretation of all the heterogeneous phenomena of nature has utterly failed. So far from reducing every type of substance or process to a single type of entity or force, the last two centuries have witnessed the discovery of numerous unsuspected forces and entities, each apparently irreducible.

The history of electro-magnetism provides the most striking example, and incidentally offers an instructive parallel to the history of parapsychology. To the earliest scientists the phenomena of lightning, of the loadstone, and of the precious substance known as 'electron' (amber) seemed so startling and fantastic that they were for long relegated to the limbo of magic, superstition, or the supernatural. Plato remarks on 'the marvellous tales about the attraction of electron and the Herculean stone', but doubts whether these substances really possess the fabulous powers ascribed to them (*Timaeus*, 80C). Apart from one or two notable eccentrics, few scientists troubled to investigate the facts at first hand until the latter part of the eighteenth century; and not until the middle of the nineteenth century did any of them have the faintest inkling of the ubiquity and the astounding potentialities of electro-magnetic forces. When at length men like Faraday, Maxwell, Kelvin, and Hertz began to study such phenomena systematically, the wildest efforts were made to 'bring them within the framework of accepted science' by reducing them to terms of mechanical processes in some quasi-material substance such as the ether.

When these attempts at unification broke down, the opposite course was adopted, and endeavours were made to derive the laws of mechanics from the equations for electro-magnetism. Once again the result was total failure; and by the beginning of the twentieth century the enthusiasm for an *Elektromagnetisches Weltbild* was on the wane. The simple planetary picture of the atom as consisting of negative charges in the form of electrons revolving round a positive charge as a centre or nucleus gave way to a far more complicated model, as more and more 'fundamental particles' were discovered. Some of the particles have an electrical charge but no mass; others have mass and no electrical charge; and many possess strangely novel properties utterly unknown to classical physics. Instead of reducing the two familiar forms of interaction—gravitational and electrical—to a single type, the nuclear physicist has been driven to recognize two more—the 'weak interaction', which is yet stronger than the gravitational by a factor of about 10^{27}, and the 'strong interaction', which is 100 times stronger than the electro-magnetic. Each type has its own special laws; and each has so far defeated every attempt to reduce it to any other type. The modern physicist would thus have no difficulty in conceiving two completely separate universes, coexisting simultaneously —the one consisting of changing configurations of particles carrying electro-magnetic energy, but, like photons, possessing no mass, and therefore incapable of gravitational interaction; the other consisting of particles possessing mass and therefore exhibiting gravitational interaction, but like the neutron (and most mesoscopic objects) possessing no electrical charge, and therefore incapable of electrical interaction. In a dualistic cosmos the two systems might interpenetrate; yet an observer belonging to one system would have no inkling of the existence of the other. In the actual universe as we know it, certain entities are subject to both kinds of interaction. Hence, for reasons we cannot guess, the two are in some degree linked, though for ages the very existence of distinct electro-magnetic forces was either unsuspected or summarily denied. Accordingly, since the physicist, whenever his observations and experiments seem to require it, does not scruple to postulate entirely novel entities and entirely novel modes of interaction, there can be no reason why the psychologist or parapsychologist should not also postulate irreducible psychic entities and irreducible psychic modes of interaction, if these help him to interpret the anomalous data with which he is faced.

(ii) But secondly, it is argued, the conception of an immaterial mind or 'psychic factor', which both in its cognitive and in its conative activities is supposed to interact with the material brain, would 'violate the laws of causal determination; and, when its cognitive activities are assumed to include such "paranormal" achievements as those of telepathy, clairvoyance, and precognition, the violation exceeds anything that the reputable scientist could possibly swallow'. However, as we have already seen, there is nothing sacrosanct about these so-called 'laws'. 'Ten years ago,' wrote Eddington in 1935, 'practically every physicist of repute was, or believed himself to be, a determinist; but it has now become clear that the old causal laws are no more than statistical laws. . . . The physicist no longer seeks to predict what will *certainly* happen in this or that individual case, but only what will *probably* happen in a group.' 'Had Gessler,' says Niels Bohr, 'ordered Tell to demonstrate his marksmanship by shooting a hydrogen atom off his son's head by means of a single alpha particle, the utmost skill in the world would have availed him nothing; whether this particular particle missed or not would have been as unpredictable as Tell's own action in response to Gessler's command.' It is much the same with the shots of the 'percipient' in any parapsychological experiment, or the guesses of a good clairvoyant or telepathist; neither they nor anyone else can tell which particular guesses will turn out to be correct. Thus, so-called 'extra-sensory perception', it would seem, is a statistical phenomenon, not a specific reaction caused by a specific quasi-sensory stimulus.

Similar arguments and counter-arguments may be adduced in the case of precognition. 'The very notion of precognition (we are told) implies that the effect precedes its cause—an assumption which every scientific thinker regards as inconceivable.' But the inconceivable has been conceived. Wheeler and Feynman, for example, are ready not only to accept action at a distance, and allow that it may occur instantaneously, but in their discussion of what are termed 'advanced potentials' they have gone so far as to suggest that the order of cause and effect might be reversed.[51] Thus, in view of the revolutionary changes in twentieth-century physics, the whole notion of causation needs to be radically revised, if not completely abandoned.[52] And in fact it would be easy to show that every one of the 'basic requirements of science', which, as the behaviourists allege, would be violated by the theories of mentalism and parapsychology are just as freely violated by the hypotheses of contemporary physics. The principle of indeterminacy is today accepted without a qualm; the so-called 'laws of Nature', as is now generally recognized, can claim validity only within a certain limited range; and the singular phenomena presented by the more distant quasars and the newly discovered quasi-stellar galaxies have led many cosmologists to

[51] *Rev. Mod. Phys.*, Vol. XVII, 1945, pp. 157 ff. For a fuller discussion of this notion, and remarks on its possible bearing on precognition, see A. E. Dummett, 'Can an Effect precede its Cause?' (*Aristot. Soc. Supp.*, Vol. XXVIII, 1954, pp. 27 ff.).

[52] The fact that some of the latest investigations appear to throw doubts at least in certain cases, on the distances formerly inferred does not affect my argument. My contention merely is that contemporary physicists are quite *prepared* to throw overboard the old doctrine of immutable laws, not that they actually *have* done so.

contend that these very 'laws of physics', which Mr. Hammond declares to be 'immutable and universal' (and therefore always the same, independently of position in space or time), may perhaps have been quite different a thousand million years ago. To a Victorian scientist the story of the universe as it unfolds itself today would have seemed a far-fetched flight of science-fiction even more preposterous than Alice's adventures in Wonderland; and, as we ourselves contemplate the new-fangled concepts of broken parity, of anti-particles and anti-matter, we seem to be following her into a veritable looking-glass universe, with the further peculiarity that if Alice met anti-Alice, or Burtt met anti-Burtt, each would at once annihilate the other.

Accordingly, when the modern physicist does not scruple to entertain hypotheses that infringe what were formerly held to be 'the universal principles of natural science', both psychologists and parapsychologists can safely claim that the antecedent improbability of their own mentalistic hypotheses is no greater than that of hypotheses which have already been adopted or advanced in realms of nuclear physics and contemporary cosmology.

4. *Physiological.* Of the remaining 'limiting principles', as formulated by Broad and Hansel, the second and third claim to express the ordinary assumptions about the relations between cognitive processes and the processes of the brain. In the more extreme forms in which they are stated by Hansel, McLeish, and Hammond they are virtually a revival of Huxley's mechanistic theory.[53] There are, however, three outstanding phenomena, which any such theory totally fails to explain, and which *prima facie* seem to need some kind of mentalistic hypothesis—(i) the existence of consciousness, (ii) the apparent efficacy of consciousness, and (iii) the peculiar unity of consciousness.

(i) The brain, as the physiologist or biochemist studies it, and as the surgeon sees it when he opens up the skull of a conscious patient, appears to be a purely material object; and its properties, like those of other material objects, consist of a characteristic physical structure, a characteristic chemical composition, and a capacity for producing certain specific chemical and electrical changes (those exhibited by all nerve-tissues when effectively

[53] Quoted above p. 102; cf. W. R. Hammond, *Brit. J. statist. Psychol.*, Vol. XIII, 1960, pp. 62 ff., who cites recent textbooks by Taylor, Bugelski, and Mandler and Kessen in his support.

stimulated). From the standpoint of the physiologist, its chief feature is its amazing complexity—over 90,000,000,000 cells arranged in a systematic fashion, all operating on a power of between 10 and 20 watts, such as would just serve to light a dim electric bulb. Now the extraordinary thing is that the physico-chemical activity of this small lump of matter is by some occult miracle of nature accompanied by consciousness; it is *aware* of what is going on around it. And much of this conscious experience shows a broad correspondence with the parts of the cerebral cortex that appear to be most active at the time. If, however, we suppose, as Huxley and most other 'materialists' in the past have supposed, that the brain itself is able to generate consciousness, then we shall be obliged to admit that this singular organ must possess properties that are possessed by no other material substance. Indeed, we could no longer regard it as a purely material entity, as the term material is ordinarily defined.

However, as I have already indicated, many psychologists and parapsychologists have maintained that the brain is an organ not for generating consciousness but rather for transmitting, limiting, and directing it.[54] When we consider the multitude of sensory impressions raining in on us, through our eyes, ears, nose, and skin, from our muscles and viscera, and the countless feelings, thoughts, and memories that are tending to surge up from below the threshold of consciousness—nearly all of which pass totally unnoticed, we realize that one of the most vital functions of the brain is to act as a selective filter, damping down and inhibiting

[54] This view is at least as old as Kant. 'The body,' he writes, 'would thus be, not the cause of our thinking, but a condition restrictive thereof, and, although essential to our animal and sensuous consciousness, may be regarded as an impeder of pure spiritual life' (*Kritik d. reinen Vernunft*, 2nd edn., p. 809). The best known statements of this theory are those of James (*Human Immortality*, London: Constable, 1898) and Bergson (*Evolution Créatrice*, 1907. Eng. trans., London: Macmillan, 1911). Their version, however, is in my view open to serious objections. I have endeavoured to suggest a modified version in my paper in 'The Structure of the Mind', *Brit. J. statist. Psychol.*, Vol. XIV, 1961, pp. 518 ff. The most eminent of living neurologists who has specialized in the study of the brain is Professor Sir John Eccles; and he gives it as his view that 'the structure of the brain suggests that it *is* the sort of machine that a "ghost" might operate', where the word 'ghost' is used to designate 'any kind of agent that defies detection by such apparatus as is used to detect physical agents' (*The Neurophysiological Basis of Mind*, London: Oxford University Press, 1953, pp. 278 ff.).

a vast mass of possible experiences, so that our attention is focused on the one thing which at the moment is essential for terrestrial survival and well-being. On this view, therefore, we should be led to infer that the material brain, so long as it is alive and operative, must be in constant interaction with some immaterial source of consciousness, which we may conveniently call a 'mind' or (to avoid associations that may prove misleading) a 'psychic' or 'psychogenic' factor. Such a conception is admittedly no more than a hypothetical conjecture. But so is the notion of energy or an electrical field. And, like them, it helps to make a wide variety of accepted facts far more intelligible. If, then, we adopt it as a reasonable working hypothesis, it would plainly reduce still further the antecedent presumption against the apparent findings of parapsychological research.

(ii) In the day-to-day business of ordinary life I find myself intuitively assuming (and I suspect every unbiased reader does the same) that this awareness, particularly this gift of selective attention, and most of all the decisions I may reach as a result of reasoning or reflection, directly affect my consequent behaviour. When I consider the conduct of other people and of other animals, I find that with them too the success with which they adjust their actions to the situations confronting them varies in close proportion to the degree with which they seem aware of things and attend to things, and the clarity with which apparently they reason out their problems. I note too that, especially in the case of animals and young children, the pleasure or satisfaction that accompanies any new and successful type of reaction helps to drive it home, so that it tends to be repeated, whereas pain, unpleasure, and the dissatisfaction that follows failure, usually inhibit the reaction. So far therefore as we can trust these intuitive impressions, we seem bound to infer that mind can influence matter, and that consciousness, by some kind of normal psychokinesis, can directly influence the activity of the brain-cells or change the direction of the neural currents.

All this, of course, no present-day behaviourist would allow. Usually the controversy is debated with special reference to the time-honoured problem of 'free will', that is, in terms of creative or purposive behaviour, i.e. conation. I would rather emphasize the apparent influence of cognitive and affective processes. What, then, are the criteria on which the behaviourist relies in attempting to refute this common-sense conception? They rest, not on any empirical evidence or proof, but almost entirely on his basic metaphysical assumptions, often so worded as to sound like statements of fact. 'Both the formal and the final aspects of that activity which we are wont to call "mental",' write McCulloch and Pitts, 'are rigorously deducible from present neurophysiology.' Taken literally this is shamelessly incorrect. No neurophysiologist has ever succeeded in making any such deduction. Hebb puts it more cautiously: 'It is unnecessary that the student of personality talks in neurological terms, but his terms should be *translatable* into neurology.' How is the student to 'translate' the *feeling* of pleasure, the *evaluation* of an ideal, or the *awareness* of the logical transition implied by the little word 'therefore' into 'neurological terms', much less 'deduce' the existence of such forms of consciousness from the sole facts of neurophysiology? As neurologists from Sherrington onwards have repeatedly insisted, 'we know far more about the mind than we do about the brain, and we understand conscious processes far better than we understand neural processes'.

For the current version of the old epiphenomenalist argument, we may take Hammond's reformulation:

> Hebb [he writes] has conclusively shown that we can infer and understand the mental processes of others only by observing their behaviour. As for our own conscious processes, they are nothing but the shadowy concomitants of the neural processes of the brain: they can no more influence the processes in the visible nerve-cells than the shadow cast by a moving car can alter its direction. The notion that consciousness can guide nerve-currents is as unthinkable as supposing it can guide the fall of well-shaken dice—a miracle which Rhine, strange to say, declares he has witnessed. All such suppositions are glaring violations of the very axioms of natural science—the laws of universal causation and of the uniformity of nature.[55]

[55] W. S. McCulloch and W. H. Pitts, *ap.* W. S. Fields and W. Abbott, 'Information Storage and Neural Control', *Bull. Math. Biophys.*, Vol. V, 1943, pp. 115–33 ff.; D. O. Hebb, 'The Role of Neurological Ideas in Psychology', *J. Personality*, Vol. XX, pp. 40 ff.; and *The Organization of Behaviour* (New York: Wiley, 1949), pp. vii ff. It may be noted that it is psychologists who commonly declare that consciousness is reducible to physiological terms, not the physiologists, just as it is psychologists who maintain that the brain itself may be compared to electronic mechanisms devised by engineers, and not the engineers. The *a priori* arguments I have cited are sometimes supplemented by references to analogies drawn from automatic reflexes, goal-

PSYCHOLOGY AND PARAPSYCHOLOGY

In reply, let us note to begin with that the argument of Hebb and Hammond commits the familiar fallacy of confusing the problem of the mind *as observed* with the problem of the mind *as observer*. It would be far more plausible to contend that our understanding of behaviour is derived from our observation of our own inner conscious processes than to maintain that our understanding of conscious processes is derived from our observation of outward behaviour. The rest of the argument is developed without any reference to the actual facts: after all, the whole of the quantum theory seemed 'unthinkable' until it was forced upon us by the data.

Shorn of the picturesque metaphors and stripped of the metaphysical postulates, the argument itself turns on just one simple question: how far, in the light of current knowledge, can we plausibly assume that the principle of strict causation, which works well enough in mechanics and classical physics, must also apply to the hypothetical neural processes which apparently underlie the observable changes in the field of consciousness? Certainly the brain which the anatomist dissects or the surgeon probes upon the operating table belongs to the order of what I

seeking missiles, and electronic computers. But all such analogies are no more than unconvincing illustrations; they furnish no proofs. Often the 'reduction' is effected by transparent verbal devices: whereas parents and schoolmasters talk of the child's happiness when rewarded and pain when chastised, Skinner insists that the psychologist should drop all mention of happiness or pain, and merely state that the child's action has been 'reinforced' (which in point of fact is *not* always the same thing). The familiar comparison of human beings to the robots of science-fiction plainly ignores the fact that the purposive activities of any actual robot (such as Grey Walter's 'tortoise' or Turing's theoretical computer) are really those introduced by the designer and the programmer. The present tendency is to invert the argument from analogy, and, dropping the traditional language of cause and effect, to substitute a theory of 'information' (manifestly a metaphor from cognition). Now it is not difficult to make rough armchair calculations of the 'bits' of information (*a*) stored and processed by an intellectually gifted thinker, and (*b*) capable of being stored and processed by the brain, as interpreted in terms of the concepts and assumptions of present-day neurophysiology. Then, on the lines of the argument about monkeys ultimately hammering out *Hamlet* by random taps on the typewriter, it is easy to show that the former far exceeds the latter. This seems to turn the antecedent probabilities definitely in favour of mind rather than mechanism. Still it might, I suppose, merely mean that the concepts and assumptions of present-day neurology are far from adequate.

have called mesoscopic or man-sized objects.[56] At this level, therefore, we may expect it to conform to the scientific principles characteristic of such objects; and, so long as the investigator is dealing with structures of the size of nerve-cells and their cell-bodies, this assumption still seems sound enough. But when we reach the minuter processes which appear most closely associated with mental activities, we find ourselves dealing, not with 'the visible nerve-cells', but with the nerve-*junctions*. Near the junction the fibre-like branches of the cells subdivide into twig-like filaments; and the filaments end in little buttons or knobs, which are applied, with no actual continuity of substance, to the cell-bodies or branches of adjacent cells. The knobs contain tiny vesicles, which, when stimulated, squirt out molecules of the chemical transmitter. The knobs themselves, which are only visible in an electron-microscope, are so minute, and the gap between knob and cell is so fine, that the processes involved must be subject, not to the laws of classical physics, but rather to those of quantum physics; and here, as we have seen, the 'basic limiting principles' no longer hold. The processes I have described imply that the transmission of nerve-impulses, unlike the transmission of information in an efficient telephone circuit or an electronic computer, must be interpreted in probabilistic rather than in deterministic terms. In certain circumstances, therefore, the brain, it would seem, could indeed act 'as a kind of detector and amplifier' (the phrase is Eccles') of the otherwise undiscernible influences exerted in cognitive, affective, and conative activities—in short, by what we should ordinarily describe as conscious or subconscious processes.[57]

(iii) There is one further problem. The most striking feature of

---

[56] I would prefer to substitute the notion of 'ground and consequent', and seek for sufficient reasons rather than sufficient causes; there is then no need to insist that the 'ground' should be chronologically prior to the 'consequent'.

[57] For more specific details I may refer to my paper on 'Factor Analysis and its Neurological Basis', *Brit. J. statist. Psychol.*, Vol. XIV, 1961, pp. 69 ff., and refs., and to J. C. Eccles, *The Physiology of the Synapse* (New York: Academic Press, 1964), esp. pp. 277 ff. Eccles suggests that the actual neural output may be determined by three factors: '(1) the micro-structure of the neural net and its functional properties', '(2) the afferent input' (including presumably feedback), and '(3) the postulated "field of extraneous influence" exerted by the agency we loosely call mind or will'.

conscious experience is its unity and continuity. Although our sensations spring from distinct sense-organs, and show strangely disparate qualities, they, and the various images and thoughts that occur to us at the same time, all form part of a single seamless whole. It is hard to see how this unity could be produced by a material brain which consists essentially of many millions of distinct and separate cells. Perhaps the most astonishing of all relevant findings is that reported in certain drastic operations on the brain. It is sometimes necessary for the surgeon to cut through the structural links between the two cerebral hemispheres (the *corpus callosum* and the anterior commissure), thus severing all direct connection between the right and left halves. Before such operations were performed it was confidently expected that, on recovering from the anaesthetic, the patient would either manifest two streams of consciousness, or else that one of his hemispheres would dominate, leaving one side of his body with its limbs cut off from consciousness and from volition. Nothing of this sort occurs. The bisection of the brain produces no bisection of personal consciousness: that still continues as a single unified stream. Actions requiring the co-operation of both hemispheres (binocular fusion, for example, and binocular perception of depth) remain apparently unimpaired. Apart from a slight lengthening of reaction times, the patient commonly shows no obvious change in mind, character, or general behaviour, and is aware of no division or limitation in his own personality or conscious experience.[58]

[58] See J. S. Wilkie, *The Science of Mind and Brain* (London: Hutchinson, 1953), pp. 133 ff. and refs., V. B. Mountcastle (ed.), *Conference on Interhemispheric Relations and Cerebral Disease*, 1965, and R. W. Sperry, 'Hemispheric Interaction and the Mind-Brain Problem', *ap.* J. C. Eccles, *Brain and Conscious Experience* (Heidelberg: Springer, 1965). The observations reported above do not of course imply that the *corpus callosum* and other commissures have no specific functions at all. What these may be is more clearly brought out in similar operations on cats and monkeys. In the earlier experiments the optic chiasma (where there is a partial decussation of the optic nerve-fibres) was also severed: this means that information from the left eye would be conveyed solely to the left hemisphere, and information from the right eye solely to the right hemisphere; and such effects are readily demonstrated. But if the commissures of the cerebral hemispheres alone are severed, quite elaborate methods of unilateral training and of bilateral testing seem needed to elicit the consequences; even then, as Sperry notes in reporting on his experiments, 'an animal with a split brain sometimes behaves as if the two hemispheres

5. *Psychological.* Of the various phenomena studied by the parapsychologist those for which empirical evidence appears strongest—telepathy, clairvoyance, and precognition—are all cognitive processes. The most helpful contribution the academic psychologist can make must therefore consist in an analysis of cognition. And here, I think, it would be agreed by almost all who have considered the problem, that cognition, which is the distinctive mark of consciousness, is essentially a *relation*.[59] Now a relation always

[59] When James, in a well-known article which behaviourists are so fond of citing ('Does Consciousness Exist?'), declared that consciousness does not exist, he was denying the existence of consciousness as a peculiar kind of *substance*. The distinction between a living, waking man and the same man in a state of absolute coma, which we express by saying the former is 'conscious', he never questioned. However, 'consciousness' in this sense denotes a quality. Here I am referring to consciousness as a process or 'act'.

The view summarized above was first developed by the Austrian school, led by Brentano (see more especially K. Twardowski, *Zur Lehre vom Inhalt und Gegenstand der Vorstellungen*, 1894). The clearest and most persuasive statement is that of Bertrand Russell in his early work, *Problems of Philosophy* (London: Oxford University Press, 1912, p. 79). In his later phenomenonalist phase, he denied the existence of any such relation. The contents of consciousness, he contended—i.e. the 'sensa' or 'sense-data' which we are ordinarily said to cognize—can exist of and by themselves, as it were, *in vacuo*. Hence there is no

were still in direct communication'. Conclusions drawn from animal experiments cannot be applied to human beings without reserve. In man, with the development of speech, one hemisphere (usually the left) tends to become dominant. And in the two patients described by Sperry some of his ingenious tests certainly suggest on first reading that, after the operation, consciousness as well as speech may have become for a while confined to the dominant hemisphere. But an impartial study of *all* the results reported makes it hard to accept this interpretation. In each of the cases it is difficult to determine how far some of the peculiarities thus revealed may have resulted from the damage inflicted by the disease or injury which the operation was designed to relieve; and it is doubtless varying effects of this nature that largely account for the different results and inferences reported by different investigators. In all of them, the most striking fact is that the patients themselves seem unaware of any division or limitation of their own consciousness and personality in consequence of the operation. Eccles, after referring to the observations of Sperry and others, and his own neurological researches, ends by declaring that 'contrary to the physicalist creed I believe that the prime reality of my experiencing self cannot be *identified* with . . . brains, neurones, nerve impulses, or even complex spatio-temporal patterns of impulses'; and decides in favour of the 'belief in a spiritual world which interpenetrates with, and yet transcends, the material world' (*The Brain and the Unity of Consciousness*, 1965, pp. 42–3).

implies at least *two* terms which are related. Hence any conscious situation must logically involve first a cognizing subject—a person, ego, or self—and secondly, a cognized object: this may be either a sensation (i.e. something sensed, such as a colour or noise) or a concrete thing that is perceived, or again an abstract conclusion reached by reasoning, or an event or situation expected or remembered. Moreover, the peculiar unity and continuity of my individual consciousness seem to imply that it is the same central factor or 'subject' that cognizes all the different sensations, images, thoughts, etc., which make up my consciousness at any one moment, and that the factor which cognized my past experience is the same as the factor which now cognizes my present experience. On certain occasions I can actually cognize my cognizing; I can say to myself, not only 'I feel this pain', but also 'I know that I am feeling this pain', or 'I remember that in the night I felt that pain'; in such a case I seem to be intuitively aware that the self who now remembers the pain and the self who actually felt it some hours ago are not two distinct selves, much less two distinct and separate 'passing thoughts' (as James at one time suggested), but one and the same self, an enduring and more or less permanent 'ego', lasting at least as long as the individual's brain—and possibly longer (a point we have yet to examine).

What then is the nature of this 'psychic factor' which I call my 'self', and which functions as the subject of the verb 'cognize'? It is not a *mere* grammatical subject. To begin with, it must possess certain distinctive properties or 'dispositional characteristics' (as the philosopher would style them). Some of these are intrinsic to its very nature, such as this ability to cognize. But it is also capable of learning, that is, of acquiring *new* cognitive dispositions. And I should hold that, besides its various cognitive capacities (i.e. capacities for knowing), it also has affective capacities (i.e. capacities for feeling pleasure, unpleasure, approval, disapproval, etc.), and conative or purposive capacities (i.e. the capacity to strive, choose, decide, and will). If it pre-existed before the individual body was conceived, it gave no signs of

---

need to introduce an act or relation of cognizing, much less any 'subject', 'ego', or 'self', to do the cognizing. Today, I fancy, few philosophers or psychologists would accept this view: for criticisms, see A. J. Ayer, 'Phenomenalism', *Proc. Aristot. Soc.*, 1947–8 (reprinted in his *Philosophical Essays*, London: Macmillan, 1963).

consciousness. Indeed, as Professor Ducasse has put it, at birth the factor seems rather to be, not so much a 'mind' (and certainly not so much a 'person') as 'a psychic germ from which a "mind" gradually develops' during its interaction with the brain.

Under normal circumstances *what* I actually cognize, and *what* direction my consciousness and my attention shall take, seem to be determined largely by certain activities of the brain itself, initiated sometimes from without, sometimes from within. These activities, as we have seen, lend themselves readily to modern methods of experimental study. As a consequence psychologists, in their discussion of perceptual cognition, tend to concentrate their inquiries almost entirely on these more obvious causal relations and ignore the cognitive. The two problems[60] are quite distinct; and the tendency to confuse the two has imported needless difficulties into the interpretation of paranormal cognition. Here I will only note that cognitive relations, so far from being reducible to causal relations (as physiological psychologists have been inclined to argue), are quite unique; and even in normal perception consciousness is able to span space, just as in memory it can span time. Hence I should consider it no great miracle if, when the inhibiting influence of the brain is partly and temporarily removed, I could directly cognize other persons' thoughts, perceive scenes and situations beyond the reach of human sense-organs, and even apprehend events that still lie hidden in the future.

Here, then, is one plausible interpretation which would allow us to accept telepathy, clairvoyance, and even precognition at their face value, and which can no longer be swept aside as outweighed by its own antecedent improbability. There is a further and slightly different way of interpreting such phenomena, which to my mind is yet more plausible, and all too frequently overlooked. If I may trust my own limited observations and experiments, the cognitive processes involved in so-called ESP would seem to resemble not perception but rather those semi-intuitive glimpses —those flashes of imaginative insight and implicit inferences, that are colloquially termed 'hunches'. In most cases, as the 'per-

[60] See 'The Concept of Consciousness', *Brit. J. Psychol.*, Vol. LIII, 1962, pp. 236 ff. It is chiefly, though not perhaps entirely, this preoccupation with the causal processes involved—particularly those studied by the physiologist— that has been responsible for the so-called 'sense-datum theory' (see below).

cipient's' introspections show, the contents of the extra-sensory experiences are not perceptual, but consist rather of mental images (often no doubt with a dream-like vividness), sudden ideas or impressions, and even of veridical utterances and impulsive decisions which the subject himself finds difficult to explain.

So far therefore my provisional conclusions may be summed up as follows. The recent advances in physical science, as far-reaching as they were unexpected, have now so drastically diminished the antecedent probability *against* so-called paranormal occurrences that it is more than counterbalanced by the high empirical probabilities in their *favour*, which emerge from the accumulated results of numerous experimental studies. At the same time, the utter failure of all purely physical hypotheses to account even for the most normal phenomena of consciousness leaves the door wide open for a psychical or mentalist hypothesis. Of all apparently paranormal phenomena that which is most strongly confirmed by experimental evidence is what is loosely termed 'telepathy'. Here, the primary process seems to be one of psychic *interaction* rather than of direct inter-psychic *cognition*. The experimental evidence for 'clairvoyance' is almost as strong; and here the cognitive process seems more direct. But so-called 'clairvoyance' appears to involve the apprehension, not of something literally 'seen', but of objects or situations which are realistically imagined or inferred. The cognitive processes involved in cases of precognition are in my view more indirect. As for the precise conditions at work, I am tempted to argue that, in the present state of knowledge, it is quite futile to seek causal explanations for paranormal processes as it is for any other type of conscious process.[61]

---

[61] Broad's detailed examination of the different types of cognitive relation that might be involved in paranormal processes (*Religion, Philosophy, and Psychical Research*, pp. 27 ff.) is an admirable example of the kind of psychological analysis that is urgently needed. But his arguments are evidently based on the assumption of the so-called 'sense-datum theory'. According to this theory, what we see when we look at a table or chair is not a physical object but a sense-datum: I hold exactly the opposite, at least under ordinary practical conditions: (the state of mind of the sophisticated psychologist, puzzling over his own experience, is in my view somewhat abnormal). I regard the isolated sense-datum (if I may echo the White Rabbit) as merely

An obstacle that comes between Him and ourselves and it.

The independence of mind

There remains, however, one still more elusive problem, namely, how far is the hypothetical 'psychic factor' or 'psychic field', defined as above, independent of the physical organism? Few psychologists, whether they accept mentalism or reject it, have realized what a wide range of possible answers there is to this basic question. The simplest form of the hypothesis would be that the psychic field or factor only functions in connection with a living and effective brain, much as a magnetic field is observable only while a live current is passing through the coil of copper wire. When the activity of the higher parts of the brain (e.g. the cortex and reticular formation) drops below a specific liminal level (as in dreamless sleep or under a general anaesthetic), then the field of consciousness subsides and ultimately seems, as it were, to evaporate altogether. And when the brain is permanently destroyed in death, then, we might infer, personal consciousness is not merely suspended but permanently extinguished. Going towards the other extreme, we have the various theories which postulate something which may be called a transcendental mind. Descartes and his followers contended that each mind or self was an eternal, independent, immaterial substance, interacting during life on earth with the material brain or some specific part of it. Other writers have carried this transcendentalism to its furthest limit, and assumed only a single universal mind or *Weltseele*, which temporarily seems to split into a number of individual minds or selves when it manifests itself through the instrumentality of individual organisms or brains.

Probably the most plausible forms of the psychic hypothesis are to be sought somewhere between these two extremes. I myself find it difficult to reconcile the phenomena, not only of parapsychology, but also of normal psychology, with the assumption that the 'psychic factor' is literally located in space. Believing as I do that ordinary space and time are merely human ways of ordering certain systematic relations, I would prefer to suppose that each psychic factor has its own specific 'mental position' in what it is convenient to treat as a fourth dimension of space: (Broad

---

But these are issues too complex and controversial to be debated here: (for a further discussion, see 'The Sense Datum Theory', *Brit. J. statistic. Psychol.*, Vol. XV, 1962, pp. 165 ff.).

regards such a position as defined by a 'positional *quality*': I should rather regard it as defined by certain *relations* between minds); and instead of supposing the 'psychic factor' to be 'timeless' or 'eternal', I would prefer to think of it as existing in a second dimension of time, which may be called 'mental time'. The introduction of these additional dimensions I regard, not as a statement about the universe, but rather as a linguistic device which enables us to formulate more precisely the numerous observable variations we encounter in normal psychology.[62] At the same time they suggest convenient hypotheses for describing certain paranormal phenomena. Just as we identify physical bodies and physical events by specifying their position in a hypothetical physical space and time, so we can identify individual 'minds' and particular mental events by locating them in these supplementary dimensions. We can then talk of the mental proximity of two minds, and describe telepathy in terms of a temporary 'mental contact'. If the 'minds' are conceived as fields, we may (if we want to) think of two or more of them as 'fusing into one' (as Abélard wrote to Héloise) or as 're-merging in the general Soul'[63] (as the mystics and panpsychists have commonly maintained). Precognition might be interpreted as due to a change in *mental* time, analogous to the apparent time-reversals envisaged by certain writers on relativity and subatomic theory. And if an individual mind survives bodily death, we could conceive it, not as passing into a kind of timeless and static eternity, nor yet as existing 'in

[62] I use the term 'dimension' rather as it is used in psychological factor analysis; in other words, what I use it to describe is not concrete 'reality' but a theoretical 'model'. In contrasting 'mental time' with 'physical time', I am reminded of a fragment from the most famous scholar of my old school:

> 'Two lovely children run an endless race—
> A sister and a brother.
> This far outstrips the other;
> Yet ever runs she with reverted face,
> And looks and listens for the boy behind,
> For he, alas, is blind!
> And knows not whether he be first or last.'

S.T. Coleridge, *Time, Real and Imaginary: An Allegory*.

[63] Tennyson, 'In Memoriam', XLVIII. In spite of all that James has urged in support of these notions of fusion or reabsorption, I myself feel that he and other champions of the same idea fail entirely to meet the criticisms set out by Pringle Patterson (*The Idea of Immortality*, Oxford: Clarendon Press, 1922, pp. 162 ff.).

some unchanging form transcending time,'[64] but as still capable, as Kant affirmed, of undergoing a progressive and creative development of its own.

We are left with one final question, that of the content of consciousness, including the puzzling problem of personal memory. The traditional conception of the mind treats it as a 'substance' characterized by consciousness, and the contents of consciousness as changing 'states' of the mind. And the ordinary notion of a person or self implies the possession of an individual history. However, the term 'substance' carries a variety of disputable associations, and raises a number of confusing side-issues. It would be better to substitute the logician's term 'continuant' (which W. E. Johnson uses to denote a relatively permanent 'existence', as distinct from other types of 'existents', e.g. 'occurrents' or 'events'), or the physicist's term 'system' (which Margenau uses to denote a 'complex construct', as distinct from 'simple' entities, if there are such things), or finally some still more general term such as 'psychic factor or component' (the designation proposed by Broad): in the preceding discussion I have preferred this last phrase, which seems as non-commital as any such designation can be. I defined it above as an interacting factor which possesses the capacity for conscious activities—cognitive, affective, and conative; but I deliberately left the further issues open.

If we adopt what is commonly known as the 'sense-datum theory' (which still appears the most popular theory among those psychologists, such as Broad himself, who do not restrict psychology solely to a study of behaviour), then it would seem most natural to infer that *what* the mind cognizes under mundane conditions is determined wholly or in part by the brain, or by a joint interaction between the brain and the psychic factor; and from this it would follow that the unique quality of an individual mind or self as well as its various acquired characteristics (memories, skills, interests, ideals, etc.) are wholly dependent on the peculiarities of the individual's physical organism and the modifications of his brain-structure. Accordingly, even if, for the sake of argument, we suppose that the psychic factor can survive the disintegration of the brain, it becomes extremely hard to conjecture what after-death experience could possibly be like. On this hypothesis it would contain neither sensations nor sensory

[64] Venkhata Rama's definition of Nirvana.

images, since these are presumed to be strictly correlated with neural processes in the sensory areas of the cortex.[65]

Nevertheless, we must remember that sense-data, though the most conspicuous contents, are by no means the only contents of consciousness, nor are habit-memories (those based on repeated associations) the only types of memory. Thoughts, meanings, specific desires, what is (as we say) 'recognized', whether in memory or in perception—these are 'contents' (if we insist on this terminology) for which no one has suggested any plausible neural correlates. And if 'extra-sensory' modes of cognition are possible in this life, I see no reason why they should not still be available for the 'psychic factor' even in a disembodied state. Moreover, convenient though it may be for purposes of exposition, I myself find it impossible to square a thoroughgoing sense-datum theory either with the introspective analysis of sense-perception or with what is now known of the neural processes within the brain. These, however, are highly speculative problems which it would be unprofitable to examine here. My own conclusion is that the question whether the hypothesis of an independent 'psychic factor' is valid or not, and, if so, what is the relative influence of each of the two co-operative 'factors', the physical and the psychical, is an issue which is to be settled not by armchair cogitation or discussion but only by *ad hoc* empirical research.

IV THE PROBLEM OF SURVIVAL

The most direct evidence bearing on the theories we have been discussing would be that which is furnished by empirical inquiries

[65] Professor Price has suggested that 'the next world might be conceived as a kind of dream-world' and the experience of a discarnate mind might be 'an experience in which imaging replaces sense-perception', since sensations depend on the effects of sensory physical stimuli on the 'brain centres'. He emphasizes that 'an *imagy* world would not necessarily be an imaginary world', i.e. unreal, but 'the joint product of a group of telepathically interacting minds' (H. H. Price, 'Survival and the Idea of Another World', *Proc. S.P.R.*, Vol. L, 1953, pp. 1–25). But images appear to be quite as dependent on 'brain-centres' as sensations: the only difference is that in 'imaging' the centres are excited from within the brain rather than from without. The rest of his arguments, particularly those which he urges against a 'completely somatocentric theory of personal identity', seem to be perfectly cogent. However, in the light of the evidence accumulated in favour of extra-sensory cognition, it would seem to me that the whole epistemological problems of perception and memory call for an intensive re-examination.

into the question of survival. During the present century, however, both parapsychologists and general psychologists have fought shy of this time-honoured problem. Yet the issue is not only of paramount importance for theoretical psychology; it is even more urgent, because of the manifest implications it carries both for practical psychology and for religion, ethics, and philosophy.

Patients who consult psychologists are, as Jung and others have repeatedly observed, often those who have 'lost their religious faith', or have become emotionally disturbed owing to conflicts about what are loosely called 'religious difficulties' (as we have seen, they are really philosophical problems, similar to those that Socrates was so fond of discussing with his disciples). In this country the general medical training tends to place paramount emphasis on a materialistic approach; and the attitude which this imparts has been still further reinforced by the thoroughgoing scepticism of Freud and his followers.[66] Thus, as numerous published and unpublished case-histories demonstrate, the attitude of the psychiatrist (inferred, it may be, from what he fails to say rather than from any overt statement) only leaves his patients still further convinced of the meaninglessness of human life. They end with the feeling which Sartre has expressed so pungently—that 'we and the universe are sheer accidents, and both are *de trop*'. Indeed, to the would-be psychotherapist, brought up on Freudian determinism and Pavlovian materialism, one is tempted to echo Jupiter's protest to Orestes in Sartre's cynical comedy: 'Pauvres gens, tu vas leur faire cadeau de la solitude et de la honte; tu leur montres soudain leur existence, leur obscène et fade existence, qui leur est donné pour rien!'[67]

In dealing with delinquents the same problem reappears in a

[66] 'Nature,' says Freud, 'rises up before us—sublime, pitiless, inexorable.' Religion, we are told, is no more than an 'obsessional neurosis of humanity', springing plainly from a baseless wish-fulfilment. 'At this point,' he adds, 'the activity of the spiritualists comes in: they are convinced of the immortality of the soul. Unfortunately they have not succeeded in disproving that the appearances and utterances of their "spirits" are merely the productions of their own mental activity' (S. Freud, *The Future of an Illusion*, Eng. trans. by W. D. Robson-Scott, Inst. of Psycho-Analysis, London, 1928, pp. 27 ff.).

[67] J. P. Sartre, *Les Mouches*, Act. III, sc. ii. And a similar view has been put forward by British popularizers of Sartre's existentialism, e.g. Colin Wilson in his earlier writings.

form still more acute. It is, and has been, widely assumed by psychologists as well as by ordinary mortals that the actions of all human beings are determined by a desire to maximize their own personal pleasure and minimize their own personal pain. Hence, it is said, if there is no prospect of reward or punishment in the future world, the mass of humanity will follow their own selfish desires, and, provided there is a reasonable chance of escaping detection, will be ready to abandon themselves to a life of sin or crime. 'The stoic attitude of a Marcus Aurelius or a Huxley,' so McDougall tells us, 'may suffice for the few who can rise to it; but I doubt whether entire nations could come up to this austere level, or even maintain a decent standard of conduct, after losing these beliefs'.[68]

Here I am not concerned to consider how far such inferences are justifiable, either logically or morally. What I wish to stress is simply this. Those who opt for the more sceptical view and hold, as Dr. McLeish and Dr. Wallis do, that 'the whole notion of a self-subsistent and self-surviving mind can be rejected out of hand as no more than a sentimental wish-fulfilment, beyond the margin of scientific credibility', have jumped a little too hastily to this sweeping conclusion without stopping to realize the extreme complexity of the issues involved. Freudians and others who talk of 'wish-fulfilment' should perhaps be reminded that, contrary to the prevailing impression, the wish for survival is not nearly so widespread as the belief.[69] Certainly the effort to survive is

[68] *Body and Mind*, p. xiv. His words are almost a verbatim echo of Renan's (*Vie de Jésus ad fin.*) Freud himself has emphasized this consequence in the monograph already quoted (*op. cit.*, p. 69). A common answer to arguments of this type, as McDougall goes on to point out, consists in defending the notion of human immortality along Kantian lines as a postulate of the 'practical reason'. The clearest modern statement of this view is that of Professor A. E. Taylor in a well-known article on 'The Moral Argument for Immortality'. I agree with Professor Broad, who examines the 'argument' in detail, that as a philosophic thesis it is far from convincing (*The Mind and its Place in Nature*, London: Routledge, 1955, Ch. XI, 'Ethical Arguments for Human Survival'). Yet, even if our ethical beliefs cannot validly be used to support a belief in human survival, there is little doubt that for many a disbelief in human survival tends to create a corresponding disbelief in traditional ethical precepts.

[69] See the results of the questionnaire on 'Human Sentiment in regard to a Future Life', issued by the American Society for Psychical Research: the statistical data are analysed in *Proc. S.P.R.* XVIII, 1903.

instinctive, as each of us realizes every time he dodges an oncoming bus. Yet there have always been a number of exceptional people who, in their heart of hearts,

> Thank whatever gods there be
> That dead men rise up never,
> That no life lives for ever,
> That even the weariest river
> Winds somewhere safe to sea.

And Professor Broad himself has declared that 'for my part I would be more annoyed than surprised should I find myself in some sense persisting after the death of my present body'.[70]

Let us then begin as before by asking first of all what is the *antecedent* probability or improbability of human survival—that is, its probability estimated on the basis of our *general* views about the universe and about human nature, before any *specific* investigation into the problem has been carried out, or at any rate before we have examined the results of such investigations. An estimate of this kind is bound to be a somewhat personal affair. Each man's assessment will vary with his temperament, his upbringing and intellectual background, his explicit or implicit metaphysical assumptions (including his religious beliefs, affirmative or negative as the case may be), and finally his knowledge of the relevant facts and scientific theories, or what are accepted as such in his immediate cultural environment.

Myers' book *Human Personality and its Survival of Bodily Death*, as its title indicates, was concerned with the arguments for the 'survival of personality after death'. We ought first, therefore, to decide what we are to understand by (i) 'personality' and (ii) by 'survival'. Take McTaggart's theory of reincarnation.

---

[70] Epicureans as well as Stoics reject 'all hankering after posthumous survival.' Cf. Lucretius, *De Rerum Naturae*, III. *ad. fin.*, who observes that, if the voice of Nature suddenly asked:

> 'Quid tibi tanto operest, mortalis, quod nimis aegris
> luctibus indulges? quid mortem congemis ac fles?'
> Quid respondemus, nisi justam intendere litem naturam
> et veram verbis exponere causam?

The Stoic view is nobly put by Epictetus in a famous passage (*Discourses*, IV, x). Marcus Aurelius's answer, if more profound, is a little more wistful (*Meditations*, XII, v). For Broad's personal view, see *Lectures on Psychical Research*, London: Routledge, 1962, last page.

Suppose that on 31 March I die as Burtt, and on 1 April wake up as the Emperor of China (the example is from Leibniz). The Emperor presumably would remember nothing of his pre-existence. If *per impossibile* he claimed to be Burtt *redivivus*, what criteria could a court of law apply to decide the issue of identity versus false impersonation? And could the posthumous existence of a mere 'psychic factor', without the memories stored in my brain, count as a satisfying form of survival? We recall how the shades of Odysseus' mother and fellow-warriors gazed on him 'with vacant and unrecognizing eyes', till they had drunk the sacrificial blood. If, being a Jew, I were to rise again in the flesh and dwell for ever in Ezekiel's New Jerusalem, or if, being a Muslim, I found myself reclining with former friends in a garden of fruits and flowers, and waited on by dark-eyed houris, all innocence and fire, I should, I imagine, still be aware of myself as Isaac Isaacson or Abou ben Adhem (or whatever my name had been), much as I should if I woke up after a night journey at a holiday resort in the Near East. Were I, however, transformed into one of the cherubs described by Lamb, or transported into a timeless and impersonal ecstasy, could I then be said to 'survive'?

Myers' wording, vague as it is, makes it clear that by the 'personality' called 'John Doe' is meant, not indeed the visible and fleshly shape—the material body previously called 'Doe', but some immaterial existent, which, even after bodily death, is still capable of consciousness and still identifiable with the immaterial existent that was presumably associated with the material body known as 'Doe' while it was alive—a mind, a self, a soul, possibly a spiritual or astral body, possibly a purely psychic entity or ego, but still possessing most of the distinctive capacities, memories, and temperamental traits that characterized the person during his incarnate existence.

For those who, like the thoroughgoing behaviourist, adopt physicalism as their metaphysical creed, the antecedent probability of survival in any such sense must be virtually zero; on their theory consciousness can be no more than a transient epiphenomenon generated by those highly complex but equally unstable collocations of matter we call brains. Dualism is manifestly far more favourable to survival, and particularly to survival in a personal form. Idealism is not incompatible with the notion of personal survival; but most idealists, as we have seen, have inclined towards some kind of panpsychic doctrine, which implies the permanence of consciousness in an impersonal rather than a personal form.[71]

In deciding which of these philosophical interpretations is least open to criticism, we must consider first and foremost the relevant findings of the different branches of natural science. But science alone, as I have already emphasized, tells us far less about the universe than is commonly supposed. We must therefore extend our survey into other fields as well. The study of human history, of literature and art, of moral, aesthetic, and mystical experience—these too may conceivably afford insight into aspects of reality and into man's place within the cosmos, which the natural sciences necessarily leave on one side.

How, then, are we to assess the antecedent probabilities of survival and non-survival so far as they can be provisionally judged from these more general considerations? Professor Broad has stated and discussed most of the commoner arguments, both *pro* and *contra*: and it is unnecessary to repeat them here. He summarizes the main results as follows. 'All in all they provide us with little more than tentative conjectures, and nothing at all that deserves the name of rational argument. The world as it presents itself in everyday experience offers no very positive reason for, and no very positive reasons against, human survival. The only reason against it is thus the utter absence of all reasons for it; and in the present case this is not a strong argument.'[72] This, I think,

---

[71] Among Christian theologians Dean Inge, if I understand him rightly, favours this wider view. 'Tennyson's oft-quoted line—"give her the ways of going on, and not to die"', he writes, 'if applied to *individual* life as we know it, seems more than we have a right to ask . . . We may look upon our lives as a temporary bubble on the water; but *we*' [and the plural is italicized] 'are the river not the bubble.' The 'revival of spiritism and necromancy, masquerading as science under the name of psychic research', he then proceeds to castigate as 'something too humiliating for jest'. The 'Platonism' which he himself advocates—the Platonism of the *Timaeus* rather than the *Phaedo*, resembling Eastern rather than Western thought'—'may', he concludes, 'be moonshine; but to those who seek a proof of immortality we must say, *Hic est aut nusquam quod quaerimus*' ('Platonism and Human Immortality', *Proc. Aristot. Soc.*, Vol. XIX, pp. 270–92).

[72] C. D. Broad, *The Mind and Its Place in Nature*, Ch. XIII, 'Empirical Arguments for Human Survival'. A recent critic commenting on this chapter and the foregoing arguments observes that, 'when the previous denizens of the animal and human world are taken into consideration, the

means that we must regard the *antecedent* probabilities as on the whole weighing decidedly against the notion of survival.

Everything therefore would seem to depend on the evidence furnished by *ad hoc* inquiries carried out by parapsychologists with the problem of human survival specifically in view. Practically all of them relate to phenomena which would be classed by the ordinary psychologist as 'abnormal'; and, owing to the recent decline of interest in the subject even among the keenest students of parapsychology, by far the most thorough investigations are still (with rare exceptions) those carried out during the first few decades of the work of the Society for Psychical Research.

1. The evidence supplied by reports of apparitions must, I think, be dismissed as wholly inconclusive. Hallucinatory experiences involving such phenomena are much commoner than is generally supposed. So far as my own inquiries go, those that occur during waking life seem usually to correspond to the type of mental imagery to which the percipient is prone: the vivid visualizer sees phantasms, and the audile hears voices. There is often some external stimulus which is misinterpreted or wrongly apperceived.[73] Except occasionally for its unusual mode of appearance and disappearance, the apparition has the natural, life-like characteristics of an actual person—not those of the semi-transparent phantom described in the traditional ghost story. Usually the percipient is in a drowsy state, or has her (or his) attention

---

[73] A striking example is Sir Walter Scott's vision of Byron, just after the poet had died. This he at first took to be a ghost; but subsequently found was due to the accidental arrangement of a plaid and cloak hanging in the dimly lighted hall at Abbotsford.

> *In the night, imagining some fear,*
> *How easy is a bush supposed a bear!*

And yet at such times we often seem to meet

> *Such shaping fantasies that apprehend*
> *More than cool reason comprehends.*
> (*A Midsummer Night's Dream*, V. i. 5-6, 21-2)

---

number of "psychic factors", all recognizable, becomes incredible; with the angels and the houris the "Garden" would surely be overcrowded'. The simplest answer to these Malthusian apprehensions about a 'population-explosion' in the spirit-world can, I suppose, be dismissed by remembering that the 'spirit-world' is not a spatial world, and therefore presumably has no finite limits!

momentarily distracted—as on entering a different room, looking up suddenly from work, etc. The underlying motivations and mechanisms are not unlike those discussed by Freud in his studies of nocturnal dreams. In both, the mechanisms at work seem more or less unconscious; and the illusory experiences seem generally to be connected with some repressed or half-repressed emotion on the part of the percipient, particularly fears, or grief at recent bereavement: like dreams, they often assume a symbolic or dramatic form. An unexpectedly large proportion are apparently veridical, or related in a significant way to some specific event—an accident, an unexpected illness, or (commonest of all) the death of the person seen. Obviously these are also the types most likely to get reported. The most plausible explanation would appear to be an unconscious telepathic communication coming from the 'agent' to the 'percipient'.[74]

2. Far more important are the so-called 'spirit messages' received through mediums and automatic writers. We can distinguish two main types—first, those in which the communication comes (usually in a more or less intelligible form and often in reply to direct questions) from a single medium or automatist, and secondly, those in which the message has to be pieced together from disjointed phrases and cryptic allusions coming almost simultaneously from several independent mediums or automatists.

(i) Since the former alone are familiar to the general public, I will take them first, and ask what value, if any, can be assigned to them. The critical investigator quickly discovers that the problem is one of unexpected complexity. The modern Sadducee's way of formulating the issue—*either* the 'spirit messages' are authentic

[74] One of the most interesting is that described by Jung in his autobiography (*Memories, Dreams, Reflections*, London, Routledge and Collins, 1963, pp. 289-90). Jung, knowing himself to be a vivid visualizer, observes that at the time he could not make up his mind whether his dead friend was 'really there' or whether the vision was 'just a fantasy'; but it turned out to be veridical. For surveys of the most important investigations see *Proc. S.P.R.*, Vol. LIII, cxci, 1960, and G. N. M. Tyrrell, *Apparitions* (2nd edn. London: Duckworth, 1953, with an introduction by Professor Price). Perhaps the most valuable results of the earlier studies were Mrs. Sidgwick's classification of the commoner sources of error, and the attempts to draw up criteria for distinguishing the better authenticated cases from reports or anecdotes that are scientifically worthless.

PSYCHOLOGY AND PARAPSYCHOLOGY

revelations *or* they must be gross impostures—is not only unduly simplified but positively misleading. To begin with, the processes involved in so-called 'mediumship' themselves present a peculiar psychological phenomenon, which still lacks intensive investigation by up-to-date methods of mental testing and the like. Four points deserve special notice.

(*a*) First, the accusation of downright dishonesty is all too often fully justified. There are very few mediums, unpaid as well as paid, who can resist the temptation to indulge in a little faking, fishing, and barefaced prevarication, when things go wrong. Hence the sweeping maxim, 'Once a fraud, always a fraud' should not be too stringently applied.[75]

(*b*) Far more frequent, however, and far more subtle are the effects of unconscious self-deception—a proclivity which even trained investigators seem at times to underestimate. The tendency to heighten one's statements so as to make them more interesting or enhance one's own importance as the subject of some memorable experience, the desire to avoid qualifications or reservations as indicative of an irresolute judgement, and above all perhaps the insistent need to adjust our observations and our recollections to fit our dominant hopes and wishes—these are all ingrained and natural tendencies of the human mind, as unconscious as they are automatic. It needs a long and arduous discipline to turn a man into an exact, objective, and truly scientific reporter.

(*c*) A curious characteristic of most mediums who specialize in 'communications from the dead' is their tendency to dramatize and impersonate; and with this there usually goes considerable histrionic skill. Even when no dead relative of the sitter is supposed to be communicating, the medium generally has one or more regular 'controls' or 'familiar spirits', who sustain a recognizable and often a grotesquely fanciful character-part of their own. One may often trace the successive steps in which the sources or the mechanisms of the medium's trance-utterances seem gradually to organize themselves into an imaginary personage, who eventually

---

[75] An academic colleague of mine, who was one of the severest opponents of 'spiritualism', because (he said) 'the deception practised by mediums has so frequently been exposed', seldom scrupled to cook his own demonstrations when the experiments illustrating his lectures on electricity and magnetism failed to work at the crucial moment.

provides himself with an impressive name—Imperator, Hiawatha, Rector, Ramakrishna, or the like—and a stereotyped greeting, which announces his presence like a signature-tune.[76] This tendency is closely related to the obvious liability of nearly all mediums to mental dissociation. They are, as a rule, highly suggestible, easily hypnotized, and prone to autohypnosis; and, like other hypnotized persons, they are then able to recover an astonishing number of trivial memories which during the waking state lie buried in their unconscious minds.

(*d*) One feature on which critics are wont to fasten is the banal and trivial nature of the ostensible 'messages', and the way they reflect the ideas, the cultural background, and the general *Sitz im Leben* (as the biblical scholar would say) of the medium and her immediate circle. Joad with some justice has observed that 'even if our departed relatives still possess souls, it would seem only too clear that they have no brains'. Nevertheless, this is the sort of muddled effect we might surely expect, if the situation is as I have tentatively suggested. A really fraudulent medium would surely do much better: she could memorize and adapt high-flown passages from some volume of sermons or moral philosophy, as indeed a good many commercial charlatans actually do. If, however, the disembodied 'psychic factor' has been recently deprived of the neurological mechanisms which supplied the sensory and verbal content of his former mundane intercourse, and is therefore compelled to depend on those of the medium, a partial failure to communicate would inevitably result; and the gropings, the frequent mind-wanderings, and the nonsensical patter that serves to bridge the gaps would then be a natural consequence. If I may amplify an analogy of G. K. Chesterton's, the condition resembles that of a reluctant patient, blind and dumb from birth, when asked to explain to an inquisitive but stone-deaf visitor, the meaning of a Beethoven symphony with the help of a well-

---

[76] The most familiar examples of this process may be noted in scriptural and apocalyptic literature. Whenever the prophets of Israel have some momentous utterance to deliver, it is authenticated by the prefatory announcement, 'Thus saith the Lord.' The pseudonymous books of Daniel and Enoch, and many chapters in Isaiah, carry names of famous personages who had no hand in their composition. As with the mediums, the object is not merely to lend the authority of a well-known name to whatever is said but also to indicate the attitude or the outlook adopted, and often to suggest clues when the utterances are cryptic or symbolic.

meaning but uneducated nurse, whose only notion of music has been gathered from her favourite 'pop records'.[77]

(ii) The evidence furnished by joint communications from independent automatists is at first sight far more impressive, particularly since in these cases the automatists, as a rule, are not professional mediums. The reader who wishes to embark on an intensive study of the subject, and play the armchair detective, will find intriguing problems ready to hand in the reports of these so-called 'cross-correspondences'. Shortly after the death of Myers, the utterances of one or two mediums and the scripts of half a dozen automatic writers seemed to indicate a new plan or scheme devised by the alleged communicators. Fragmentary messages or allusions, pointless or meaningless in themselves, were received independently and about the same time—all ostensibly from the same communicating spirit (usually Myers, Sidgwick, or Gurney, the founding fathers of the Society for Psychical

[77] For an illuminating discussion of the processes that might conceivably be involved in these indirect modes of communication (derived partly from the remarks of the mediums themselves and partly from the alleged accounts of the supposed communicators), see Broad *Lectures on Psychical Research*, esp. Ch. XIII.

Since the above was written a remarkable series of automatic scripts has been published, which, as Professor Broad says in his Foreword, 'form a very important addition to the material which, *prima facie*, suggests rather strongly that certain persons may have survived the death of their bodies' (Geraldine Cummins, *Swan on a Black Sea*, London: Routledge, 1965). Miss Cummins, daughter of a Professor of Medicine in the National University, Ireland, received through Mr. Salter (for long Secretary of the S.P.R.) a request from a Major Tennant, asking if she could get in touch with his mother who had recently died. The outcome was a sheaf of forty-four scripts, the first five of which gave the communicator's name as 'Win', and said she had lived at 'Cadox Lodge, my husband's family home', and later added that she had 'insisted' on being called 'Mrs. Wills . . . Not quite right'. Messages were also sent from 'Gerald'. Mrs. Winifred Coombe-Tennant had worked with the late Lord (Gerald) Balfour, and, under the pseudonym of 'Mrs. Willett', had herself published automatic scripts. This fact, a closely guarded secret, was unknown even to her family at the date of the earlier scripts. Miss Cummins, however, had for many years lodged when in England with her friend Miss Gibbes, herself a keen member of the S.P.R. To the sceptical reader this suggests a possible source of leakage, conscious or unconscious. Nevertheless, the scripts also contain numerous intimate details, subsequently verified by references to Mrs. Coombe-Tennant's diaries, which it is difficult to explain away in this fashion. Thus, as usual, the evidence though strongly suggestive, can still not be regarded as conclusive.

Research), and the various enigmatic messages, when fitted together like a verbal jig-saw puzzle, often made surprisingly clear sense. Most of the participants were educated people—Mrs. Verrall, for example, the wife of a Cambridge professor, Dame Edith Lyttelton, Mrs. Fleming, a sister of Rudyard Kipling who lived in India, and Mrs. Coombe-Tennant, a lady active in Welsh affairs, best known perhaps as the first woman appointed by the British Government as a delegate to the Assembly of the United Nations. A systematic analysis of the scripts was undertaken by the Secretary and ex-Secretary of the Society for Psychical Research, with the help of Sir Oliver Lodge and Gerald Balfour (later the second Earl of Balfour) an expert classical scholar.

The printed reports extract the relevant details—a formidable mass of data—and supplement them with a verbal interpretation.[78] Most readers, I find, in trying to form an impressionistic judgement, are in the end swayed in one direction or the other by their own personal prepossessions: one can guess their final verdict before they have read the discussion. The only way to thread one's path impartially through the complex tangle of material is first to make a list of the significant items which are independent of each other, and then to allot to each a rough weighting to indicate the probability that it has been contributed by the ostensible communicator as compared with the probability that its apparent relevance is the result of some other factor—a normal source of information, some paranormal source (telepathy or clairvoyance), conscious or unconscious deception, the misplaced ingenuity of the interpreter, or sheer chance—always taking care to give the benefit of the doubt to these more humdrum explanations. Finally, one can combine the separate probabilities by means of the recognized formulae for the combination of probabilities. Since probabilities are not added but multiplied, the cumulative figure often mounts up with an amazing rapidity. By way of a control experiment I recommend that the reader, having first applied this

[78] H. F. Saltmarsh's short survey of the earlier material (*Evidence of Survival from Cross-Correspondence*, London: G. Bell, 1938), is now, I understand, out of print. There are three excellent chapters on the subject in Mrs. Heywood's book (*The Sixth Sense*, London: Chatto, 1959). For the ordinary reader the most interesting example of all is perhaps the more recent study by the Countess of Balfour, entitled *The Palm Sunday Case: New Light on an Old Love Story*—that of the first Lord Balfour, the former Prime Minister (*Proc. S.P.R.* Vol. LII, 1960).

method to some short example—I would suggest the so-called 'Lethe Case' (*Proc. S.P.R.*, Vol. XXIV, 1910, pp. 86 ff., and Vol. XXV, 1911, pp. 113 f.)—should then put the test question—'What does the word "Lethe" suggest to you?'—to several friends who are as ignorant of classical mythology as Mrs. Piper was: he should then endeavour to link their several answers together (as is done in the children's game of 'Consequences'), and see how far he gets to a plausible interpretation like that presumably intended by the discarnate Myers.

What inferences can be drawn from the analyses of all these data? First, I think, we can without hesitation accept the conclusion reached by everyone who has troubled to scrutinize the vast accumulation of material now available—namely, that, whether or not we take the ostensible communications at their face value, they unquestionably present strong evidence for *some* kind of paranormal knowledge: the only question is—what kind? At the moment, it would seem, the hypothesis most widely favoured is that they are (if I may coin such a phrase) the result of 'normal paranormal powers', i.e. those indicated by experimental research. Again and again (so it is readily admitted) detailed items of information which the medium could not possibly have discovered by ordinary means, honest or dishonest, have been checked and found correct. But in that case, it is said, if the fact referred to is some past or present circumstance, there must either be some extant record of it, or else some person still alive, who is able to vouch for it; and consequently the relevant data would in theory be accessible to a medium with the requisite clairvoyant or telepathic powers; if the fact related to some future occurrence, then the prediction could be explained by precognition. Horatio's rejoinder therefore still holds good: 'It needs no ghost, my lord, come from the grave to tell us this.'

Accordingly, several notable critics of the 'spiritualist hypothesis' have argued that an exceptionally gifted medium, such as Mrs. Piper or Mrs. Willett, might have been able on occasion to search out the requisite information 'even though it was hidden on the shelves of unknown libraries or lodged in the minds of utter strangers'. 'Cross-correspondences' could be likewise accounted for in much the same fashion. We have only to suppose that one or more of the automatists could 'tap the unconscious minds not only of her sitters, but also of the other automatists'.

One commentator has even suggested that, all unwittingly. Mrs. Verrall herself might have 'acted as the master-mind, and conceived the whole scheme as a means of satisfying her enthusiastic interest.'[79]

Thus, as a result of its own successes parapsychology would seem to have blocked every hope of reaching a definite solution to the one problem which, more than any other, inspired its early pioneers. It is largely the realization of this apparent deadlock that has brought about the recent loss of interest in speculations about survival. But a moment's reflection is sufficient to show that any such omnibus theory involves assumptions which strain credulity to the very limit; it is one of those ingenious skeleton-keys that open every lock, and consequently provides a genuine fit to none. It pushes the notions of telepathy, clairvoyance, and precognition far beyond the processes they denote in experimental studies or objective reports. No medium has ever manifested, or indeed even claimed, such superlative powers.

However, suppose for the sake of argument that, thanks to some transcendent gift of second sight, automatists like Mrs. Piper or Mrs. Willett *were* able to 'see' the printed pages of certain out-of-the-way Greek and Latin texts to which the scripts refer. Since both were ignorant of the classical languages, they could not have understood what their paranormal vision revealed; much less could they, in response to some unexpected query from a sitter, promptly hunt up the relevant passage or phrase. What I find so incredible in the theory are not merely the exceptional paranormal powers or even the scholarly knowledge of the classics, but the exceptional flair for discovery and selection, which the theory in this form tacitly postulates.

Still, it may be said, the world must contain several well-read scholars acquainted with the works in question. Hence in virtue of their telepathic powers such mediums might still be able to 'tap the unconscious minds' of one or other of these erudite authorities. But here we have another overworked phrase that is all too glibly invoked—a phrase which itself rests on a misleading metaphor. The mind is not a storehouse or cistern which can be 'tapped' at will, nor yet (to use another popular analogy) a well-organized filing-system containing all the owner's detailed

[79] See E. R. Dodds, 'Why I do not believe in Survival', *Proc. S.P.R.*, Vol. XLII, pp. 167 ff.

knowledge pigeon-holed for ready reference. And surely the behaviourist should be the last person to talk about 'minds' in any endeavour to rebut the notion of survival. The process of recall, whether intentional or spontaneous, is an active process, depending on the re-excitement of certain living neuronic systems within the brain. Telepathy itself, in every well-established case, appears to involve an *activity* on the part of the agent who transmits the information, as well as on the part of the recipient who picks it up. The initiating process may be unconscious; but it is essentially an active, not a passive affair. So-called 'thought-reading' is not just an 'extra-sensory perception' of some 'idea', lying inertly in the informant's mind, waiting to be read off by the automatist. And once again, it must be added, no telepathist has ever been discovered with anything like this remarkable talent for seeking out the right informant and selecting the right piece of information.

The whole theory of 'super-ESP' is manifestly so far-fetched that an unbiased reader can hardly help wondering what can have driven contemporary parapsychologists to improvise an alternative explanation at once so improbable and so ill-attested in preference to the much simpler hypothesis adopted by the earlier investigators, namely, the post-mortem survival of an individual mind.[80] The reason is to be found, not in the results of any fresh researches carried out by parapsychologists themselves, but rather in the fact that their colleagues in the field of general psychology and allied branches of study—psychiatry, neurology, molecular biology, and particularly philosophy—have come more and more to the conclusion that the world is essentially a physical world, and that there is no mind to survive, that dead men tell no tales for the very plain reason that when they are dead they no longer exist. The change has thus been a change in what are

[80] The marked shift in the views and interests of leading members of the Society for Psychical Research at the present day when compared with those of the original founders has more than once been noted by writers who are familiar with the history of the Society. Mrs. Heywood, for example, begins her discussion of the subject by remarking that 'the problem of man's survival was more vivid and immediate to the early researchers than it is to us'; and she goes on to cite such eminent names as those of Myers, Gurney, Mrs. Sidgwick, Sir William Barrett, Sir William Crookes, Sir Oliver Lodge, and Lord Rayleigh as active investigators who 'in the end all accepted survival' (*The Sixth Sense*, pp. 53–5).

tacitly assumed to be the antecedent probabilities of survival and non-survival respectively, which has overshadowed the more direct evidence. Nevertheless, as I have intimated above, the pendulum seems already to have started swinging once again in the opposite direction.

The present position is admirably expressed by two eminent authorities, each of whom has not only personally subjected the relevant material to a detailed and critical scrutiny, but also carried out first-hand investigations of his own—one a leading British philosopher, the other a leading American psychologist. First, the verdict of Professor Broad:

> Mediums and their alleged 'controls' and 'ostensible communications' [he says] display a knowledge of facts about the past lives of dead persons, and the present actions and thoughts of living ones, too extensive and detailed to be ascribed to chance coincidence, and quite inexplicable by reference to the normal sources open to the medium. These gems of correct information are nearly always embedded in an immense matrix of twaddle and irrelevance. But if one confines attention to the gems, it is often hard to resist the conviction that the spirit of a certain person has survived and is communicating. If on the other hand one concentrates on the matrix, and considers the immense antecedent improbability of a human personality surviving after death, it is hard to believe anything of the kind.[81]

Next, Professor Gardner Murphy, writing from the very home of the behaviourist movement:

> As the biological evidence comes in, decade by decade, I cannot find [he writes] any easy way to conceive of a spiritual entity independent of the living system known to biology, psychiatry, and psychology. . . . Yet, struggle though I have done as psychologist for forty-five years to find a 'naturalistic' way of handling this material, I cannot do so even when using all we know of human chicanery, and the telepathic and clairvoyant abilities of some gifted sensitives; the case looks like communicating with the deceased. . . . To me the evidence cannot be by-passed; nor yet can conviction be achieved.[82]

At the moment, therefore, and I suspect for some time to come, the net result would seem to be that the antecedent or *a priori*

---

[81] *Lectures on Psychical Research* (1962), p. 259 (slightly abridged.)
[82] *Challenge of Psychical Research* (New York: Harper & Row, 1961), p. 27.

probabilities *against* survival and the *a posteriori* probabilities *in favour* of it roughly balance out, and the arguments in support of *both* alternatives are mainly negative. On the one hand, our general knowledge of brain-processes and mental processes affords little or no direct evidence for assuming a mind or psychic factor that could survive the dissolution of the brain; on the other hand, the specialized study of what purport to be communications from the dead has so far failed to supply any plausible explanation of the facts reported without introducing some such questionable assumption.

Thus, when all is said, the only conclusion we can honestly draw is that, so far as the scientific evidence goes, personal survival is a possible, but by no means an established or even (so many would say) a plausible hypothesis. Time after time, when one has hit upon what seems a crucial case, a further search betrays some unexpected loophole. And my own impression tallies exactly with that of William James. 'I confess,' he writes, 'I have sometimes been tempted to believe that the Creator has eternally intended this department of nature *to remain baffling*' [83] (his italics). This may sound disappointing to several eager supporters of psychical research who, like the more optimistic of the early founders, 'have felt assured that the collection and examination of the facts in a frankly scientific spirit will yield unquestionable evidence that there *is* a future life, that virtue *is* rewarded, and that no honest man need now fear death either for himself or for those he loves'.

And yet, if one stands back and takes a broader philosophic view, is not this non-committal outcome just what we ought to expect and desire? Such 'unquestionable evidence' would mean the end rather than the revival of a pure disinterested virtue. Jewish, Christian, and pagan moralists are agreed on two main

---

[83] 'Final Impressions of a Psychical Researcher', *Memories and Studies*, London: Longmans, 1911); reprinted as 'The Last Report' in *William James on Psychical Research*, ed. by Gardner Murphy and R. O. Ballou (London: Chatto, 1964, p. 332). James, we are told, set out to look for 'knock-down-cases', and failed to find a single one: any attempt to draw conclusions must be based solely on a balance of probabilities. Like Darwin, James, in his fixed resolve to avoid subjective bias, conscious or unconscious, not only insists on 'the rule of scientific logic never to assume an unknown agency whenever there may be a known one', but even prefers to err by giving excessive weight to the weakest points in the evidence.

points: first, that true human morality must be this-worldly and not other-worldly; and secondly, that it must be, as the scholastics used to say, essentially alio-centric not ego-centric, or, in plain terms, that it demands an utter self-forgetfulness. Clearly an interest in, and a reliance on, the survival of one's self in a blissful Utopia where we shall all be duly indemnified for any suffering or self-sacrifice we may have incurred on earth, implies, not self-forgetfulness, but a sophisticated type of sheer self-centredness: it is like Miss Beatitude's demand that 'Shakespeare's tragedies should have followed the example of the Book of Job and finished with a happy-ever-after ending'. 'Where there is no uncertainty, there can be no real heroism, no altruism, no virtue for virtue's sake.'[84]

The uncertainty leaves the question open in *both* directions. On the one hand, the theoretical psychologist (and that includes the parapsychologist) should, on this particular issue, preserve a strict agnosticism, pressing physicalistic interpretations as far as they will go, and, even if in the end he feels compelled to adopt the hypothesis of a surviving mind, he must remember that it is, like the ether of old, no more than a hypothesis. On the other hand, those who, from reasons of faith, metaphysics, or what they take to be personal revelation, still wish to believe in survival for themselves or those they love, need have no grounds for fearing scientific censure. Thus our verdict on the whole matter must be the same as that pronounced by Plato two thousand years ago—the reply he puts into the mouth of Socrates while waiting to drink the hemlock. 'I would not positively assert that I shall join the company of those good men who have already departed from this life; but I cherish a good hope.'[85] Hope implies, not the vir-

---

[84] I am quoting or rather condensing the arguments advanced by F. H. Bradley (*Ethical Studies*, London: Oxford University Press, 1876, Essay II) in criticism of the conclusion reached by Henry Sidgwick at the end of his *Methods of Ethics*, (London and Cambridge, 1874, pp. 501 ff).

[85] *Phaedo*, 63B f. Again and again throughout the dialogue he refers to immortality as 'a goodly venture' and 'a grand hope' ($\kappa\alpha\lambda\grave{o}s$ $\acute{o}$ $\kappa\acute{\iota}\nu\delta\upsilon\nu\sigma s$, $\acute{\eta}$ $\grave{\epsilon}\lambda\pi\grave{\iota}s$ $\mu\epsilon\gamma\acute{\alpha}\lambda\eta$). In the past, when I have expressed some such opinion, several friendly critics have objected that I 'appear to be denying the essential tenets of religion'. If by this they mean the doctrines of the Christian creed, let me remind them that the formula is 'Credo in', i.e. 'I put my faith *in*' such and such an affirmation, not 'I am convinced *that*' the affirmation is true. The teaching of Jesus himself seems clearly to have been in keeping

tual certainty of success but the possibility of success. And it is, I think, one important result of recent psychological and parapsychological investigations to have demonstrated, in the face of the confident denials of the materialists and the behaviourists, *at least the possibility* of survival in some form or other, though not necessarily in the form depicted by traditional piety or fourth-century metaphysics.

> ... The rest
> From man or angel the Great Architect
> Did wisely to conceal.
>
> *Paradise Lost*, VIII. 73-5

Unfortunately I wrote this chapter before the publication of *Brain and Mind*[86] otherwise I would have cited it at several points. However I would like to take this opportunity to record my agreement with the standpoint that Dr. Smythies takes up in his essay in this book.

[86] *Brain and Mind* (J. R. Smythies, Ed.). London, Routledge and Kegan Paul, 1966.

---

with what I have stated in the text (cf. Mark XII. 27; Luke XVI. 30-1). St. Paul almost echoes the words of Socrates: 'we are saved by hope; and hope that is seen would not be hope' (Rom. VIII. 24). As for other religions I need quote only one well-known saying about the Buddha: ' "Why has the Enlightened One not declared whether the saints shall live after death?" "Because, Brother, this is a matter that is not among the things needful for the holy life, nor for enlightenment, passionlessness, or peace" ' (*Samyutta Nikaya*, II. 223).

VI

BIOLOGY AND ESP

Alister Hardy

For some years now I have been trying to point out just why I believe that psychical research, with the establishment of ESP as a reality, must have a profound effect upon biological theory. I surprised many of my colleagues when I introduced the subject of telepathy into my Presidential Address to the Zoology Section of the British Association at its meeting in Newcastle in 1949. 'It is, perhaps, unorthodox,' I said 'for a zoologist to introduce such a topic, but I do so for a reason. If telepathy has been established, as I believe it has, then such a revolutionary discovery should make us keep our minds open to the possibility that there may be so much more in living things and their evolution than our science has hitherto led us to expect.' In the short time that I was then able to devote to the subject I failed, I think, to convey what I really had in mind.

I developed the theme a little farther in two papers, one in 1950 and the other in 1953, in the *Journal* and *Proceedings* (respectively) *of the Society for Psychical Research* and I have returned to it again in one of my Gifford Lectures which has recently been published with the others of the set under the title of *The Living Stream*.[1] I welcome the opportunity, which the invitation to contribute to this volume now gives me, of writing a more general statement of why I regard psychical research to be important for the future development of biology. Much of what I am saying closely follows part of the lecture I have just mentioned.

I will not take up space in reviewing the evidence for the existence of ESP of the telepathic kind. It will be seen from the

[1] London: Collins, 1965.

quotation given above that I regard its reality as already established; and I have discussed the kind of evidence which has convinced me, giving references to the original work, in the papers to which I have referred. I want here to deal with the various ways in which I believe the results of this research may have profound biological implications. If some biologists are not yet convinced, then I believe that, if they would devote the necessary time to make a reasonable study of the literature, particularly the volumes of both the *Journal* and *Proceedings of the S.P.R.*, they must at least come to the conclusion that there is sufficient evidence to point to the strongest desirability of further research to establish beyond question whether the alleged phenomena are real or not. Most biologists, of course, have such full programmes of more orthodox work in front of them that they resent the idea that they should turn aside from their preoccupations to make an assessment as to the possible validity of something they do not believe in. It is my purpose in this article to explain just why I think biologists should give more attention to these matters.

The discovery that individuals are somehow in communication with one another by extra-sensory means is, if true, undoubtedly one of the most revolutionary discoveries ever made. It is a biological phenomenon, if true, almost as fundamental as that of gravity between material bodies. So revolutionary is it that many scientists deny its possibility because it appears so contrary to their current ideas, just as the opponents of Galileo denied the possible validity of his findings on similar grounds. Such a faculty could hardly be peculiar to a relatively small proportion of the population; is it not more likely that only a relatively few individuals, and then only on certain rare occasions, are conscious of what is really a fundamental (but usually unconscious) principle of general occurrence? Again, are the features which separate man from the rest of the animal kingdom of such a nature as to suggest that this extraordinary gift is likely to be peculiar to Man alone? When we consider this relationship I think we must admit that it is more probable that such a faculty, if true, must be some general, but at present unrecognized, element in the animal world.

If true, if true, if true, we keep saying; is it not time renewed efforts were made to establish beyond any possible doubt the truth or otherwise of this so far elusive phenomenon? For me, at any rate, one of the most interesting aspects of telepathy lies in the possibility that it may be the clue to some more fundamental biological principle.

By way of introduction to what I want to say let me first stress how very close man is, in spite of all his new qualities, to the rest of the animal world. His emergence has only come about in very recent geological time, within the Pleistocene period, a mere million years compared with some two thousand million years of organic evolution; the period of modern Man, *homo sapiens*, is less than a hundred thousand years, and, of course, that of civilized Man, far, far less. Evolution merges into history; we are now in what Sir Julian Huxley calls the new psycho-social stage of evolution. This new phase, as he points out, is no longer proceeding just by the transference of the physical genetical material—the hereditary genes—but by the handing on and development of ideas. Acquired knowledge is continually being passed on; at first it was transmitted in speech, then in writing, next in printed books, and now in all manner of new means of communication. It is constantly being added to by experience; we see emerging a new element in the evolutionary stream—the verbal inheritance of acquired experience and ideas.

Evolution has passed into a phase which, if not as striking an innovation as the beginning of life itself, is a step in its history at least as great as the separation of animals from plants. As Sir Peter Medawar so vividly emphasized in the last of his Reith Lectures,[2] it has changed the main system of heredity from a Darwinian to a Lamarckian one; it is the body of *acquired* knowledge, handed on from generation to generation, always drawing new material from the environment as it goes, which is now more important than the DNA genetical system. While the change from occasional tool-using to systematic tool-making may well have preceded speech, I cannot but regard the development of language as the leading factor which has brought about the separation of man from his animal ancestors. It gave rise to reasoning, the exchange of ideas, and the initiation and spread of inventions in the art of tool-making; the growth of these activities and new ways of life, which they brought about, have, through selection, led to a greater and greater enlargement of the lobes of the cerebral cortex of the brain which are the centres of association. Revolutionary as this

[2] *The Future of Man* (London: Methuen, 1960).

development has been, there is nothing in it to lead one to expect anything like the emergence of extra-sensory perception as some new faculty of man; it might, as has indeed been suggested by some authors, notably the late G. N. M. Tyrrell in his book *Homo Faber*, have had the very reverse effect of repressing a possibly more primitive means of telepathic communication in favour of the new linguistic method which can transmit more precise information. Be that as it may, there is another point which should be taken into consideration in discussing man's relation to his animal ancestors.

It is now being realized that Man, on his emotional side, is much closer to the rest of the animal world than had been thought even by the most confirmed evolutionists. An important contribution to this change is the thesis presented by Professor Polanyi as part of his Gifford Lectures delivered in Aberdeen in 1951-2 and published, much elaborated, under the title of *Personal Knowledge* in 1958;[3] a little later he condensed the essence of his views into a much shorter book, embodying his Lindsay Lectures, *The Study of Man*[4] He points out that there are two kinds of human knowledge:

 (i) *Explicit knowledge*: that formulated in words, maps, mathematical formulae, etc.; and

 (ii) *Tacit knowledge*: unformulated knowledge, i.e. the knowledge of what we are actually in the act of doing before it is expressed in words or symbols.

The more primitive forms of human knowing—those forms of intelligence which Man shares with the animals—are situated behind the barrier of language. Babies up to eighteen months or so are said to be not much superior to chimpanzees of the same age; only when they learn to speak do they leave the apes far behind. Even adult humans, however, show no distinctly greater intelligence than animals so long as their minds work unaided by language. In the absence of linguistic clues Man sees things, hears things, feels things, moves about, explores his surroundings, and gets to know his way about very much as animals do. Polanyi quotes Tolman as saying that a rat gets to know its way about a maze as if it had acquired a mental map of it; observations on human subjects suggest that Man, however intelligent, is no better at maze running than a rat, unless assisted by notes, whether they are remembered verbally or sketched out in a drawing.

[3] London: Routledge, 1958.   [4] London: Routledge, 1959.

The sudden development of Man's reasoning mind, so recent, in relation to the millions of years of animal evolution, has resulted from the coming of speech and the making possible of explicit knowledge. In our tacit knowledge we are one with the animals. In fact, not only just one with the animals: the tacit powers of the animals, as Polanyi suggests, may be *greater* than our own. The development of our explicit powers may have to some extent dulled those at the tacit level.

> It is of course impossible [writes Polanyi] to compare exactly the level of tacit performances involved in the works of human genius, with the feats of animals or infants. But we may recall the case of Clever Hans, the horse whose powers of observation far exceeded those of a whole array of scientific investigators. They believed the animal was solving problems set out on a blackboard in front of it, while it was actually taking its clues for correct answers by watching the involuntary gestures made by the scientists themselves in expectation of these answers. Remember also how readily and how well children learn to read and write, compared with hitherto illiterate adults. There is enough evidence here to suggest that the highest tacit powers of an adult may not exceed, and perhaps actually fall short of those of an animal or an infant, so that the adult's incomparably greater performances are to be ascribed predominantly to his superior cultural equipment. Genius seems to consist in the power of applying the originality of youth to the experience of maturity.

The explicit statements of knowledge and the culture to which they have given rise, while so important in the future evolutionary process, can hardly have introduced some fundamentally different principle that was not present in the universe before them. It is most unlikely, I believe, that this new linguistic development of the human mind should have produced either the faculty of extra-sensory perception, or, as some biologists seem to imply, the state of consciousness. I agree, of course, that the complex human personality has indeed been built up in some such manner as the psychologists suggest, although the advocates of the different schools still differ very much among themselves as to the exact nature of the process. Where consciousness came in the course of evolution we have no idea. To talk of consciousness as a biological problem is today as much a taboo in orthodox scientific circles as it is to talk of telepathy. To suppose that man alone is conscious may be a comforting thought, but I believe it to be very unlikely; records like those of Mrs. Joy Adamson's friendship with her lioness in her book *Born Free* (1960) or of Miss Len Howard's studies of *Birds as Individuals* (1952) show that we have almost as much ground for regarding such animals as conscious as we have for judging our fellow-men to have this quality.

Sir Charles Sherrington, in a memorable passage in his Gifford Lectures, *Man on His Nature*,[5] says:

> We have, it seems to me, to admit that energy and mind are phenomena of two categories. . . . Mind as attaching to any unicellular life would seem to me to be unrecognizable to observation; but I would not feel that permits me to affirm it is not there. Indeed, I would think, that since mind appears in the developing soma[6] that amounts to showing that it is potential in the ovum (and sperm) from which the soma sprang.

I was delighted that Sir Cyril Hinshelwood gave such prominence to the subject of consciousness in his Presidential Address to the Royal Society in 1959. I quote from him as follows:

> The question, then, of the relation of the internal and the external worlds cannot and should not be ignored by men of science.
>
> It is surprising that biological discussions often underestimate human consciousness as a fundamental experimental datum. In science we attach no value to unverifiable deductions, or to empty qualitative statements, but nobody defends the neglect of experimental data. Among these we cannot validly disregard those of our own consciousness except by a deliberate abstraction for which we must assume responsibility, and which we should not forget having made. . . .
>
> The approach to the problem of consciousness must admittedly be egocentric. I have to begin logically with the, to me, indisputable fact that I have a wide range of conscious experiences (which are indeed more important to me than anything else). The significance of this fact can be questioned in two ways: first, by the objection that as far as any operational test goes other people may well be behaving simply as machines, and secondly, by the assertion that my consciousness is a more or less irrelevant concomitant of behaviouristic reactions.

[5] London: Cambridge University Press, 1951.
[6] For those who are not biologists I should explain that the term *soma* means the body as distinct from the reproductive germ cells.

I am, it is true, not certain that other people have consciousness. Neither indeed am I certain that the atomic nucleus exists. But I regard its existence as in the highest degree probable. I regard consciousness in other people as equally probable and for similar reasons. The atomic nucleus is not directly observed. It is inferred by elaborate reasoning from many complex experiments, any one of which could probably be given alternative interpretations. What carries conviction is the fact that a coherent body of doctrine emerges from large numbers of varied tests.

We are all the time trying experiments on our relations with other people, informing, asking, ordering, obeying, resisting with varying emotions which are correlated with our actions. Human evolution has developed an elaborate and sensitive communication system, brought into play in innumerable ways. The hypothesis that other people have an interior life not unlike my own enables me to register correspondences at point after point in so intricate a way that I accept the hypothesis, if not as absolute truth, then as something nearly as good. And with most of the basic conclusions of science I am in no position to demand more. . . .

After much more discussion of the problem he continues:

There is at present no obvious answer to the question of what kind of advance can possibly be hoped for in the problem of psychophysical concomitance. This, however, is no reason for giving up thought which at least helps to avoid the kind of errors so easily made both about physics and about biology when the problem is ignored. . . .

Human knowledge will not be in a satisfactory state until the dichotomy of the internal and the external is somehow removed.

Nothing could be more important for a true biology, one that embraces the whole of life, than a better understanding of the mind–body relationship. Orthodox biology of today is, in its mechanistic materialism, as narrow and dogmatic as was the medieval church, but in the opposite direction. A greater effort in the study of extra-sensory perception is likely to throw more light on this problem of psychophysical concomitance.

Leaving, however, on one side this greatest of all philosophical riddles, which is not likely to be solved in the immediate future, just how do I believe that a better understanding of extra-sensory perception may have an important biological significance? When

ALISTER HARDY

I introduced the subject of telepathy into my address to the British Association in 1949 those of my audience who were not entirely sceptical as to its existence either thought that it could have nothing to do with zoology or that at most it might perhaps explain certain aspects of animal behaviour such as the sudden change of direction of a flock of birds in flight as had been suggested by several naturalists, particularly by Edmond Selous in his *Thought-transference, or What, in Birds?* (1931). I was really thinking of something more fundamental from an evolutionary point of view than the closely correlated movements within flocks of birds or shoals of fish, or perhaps the explanation of the remarkable homing behaviour exhibited by some animals. Before, however, passing to the more basic considerations, I should say a word or two about these other possibilities.

While I think it possible that the movement of animals in unison may be brought about by some telepathic communication we must not jump to the conclusion that it is so; the reactions of some animals may be so much faster than our own that, if there were a wave of altered movement passing through a flock from some leader, it might be so swift as to appear to us as an instantaneous effect. With the use of high-speed ciné-photography and a close and systematic scrutiny of a series of consecutive pictures it should be possible to detect the spreading of such a wave of changing direction if it occurred; I do not, however, know of any study having been made on these lines.

Regarding the homing of animals, there are many recorded anecdotes of dogs and cats which have found their way home after being transported long distances by car or rail, and the remarkable achievements of birds such as pigeons are well authenticated. While the possibility of the operation of some form of ESP should be kept in mind, we have no sound evidence that such a method is in fact employed; experiments like those of Dr. Pratt and his colleagues at Duke University carried out to test this hypothesis with pigeons have not yet yielded any conclusive results. The problem of homing is still an open question and is more difficult to solve than that of migration; it has been well demonstrated that some species of migrating birds navigate, no doubt unconsciously, by using the position of the sun by day while other species, flying by night, use the pattern of the stars. The 'night sky navigation' has been convincingly proved by observing the behaviour of such birds in a planetarium when the 'sky' has been turned through 90° or 180°.

There are anecdotes suggesting a telepathic rapport between humans and their animals, particularly dogs and horses, as when Rider Haggard had a vivid dream that his dog was badly hurt and dying, and then found it had been killed the same night by being struck by a train; or there was the lady who suddenly felt something had happened to her favourite horse and on hurrying home found it had fallen and broken a leg in its stable. From time to time dogs have been shown off as having developed an extraordinary power of answering the thoughts of their masters; on careful investigation, however, they have almost invariably been shown to be taking clues from facial expressions and other unintentionally expressed signs, as did the celebrated horse Clever Hans to which Professor Polanyi referred in the quotation on p. 147. The alleged powers of the dog Chris investigated by Dr. Pratt and other members of the Parapsychological Laboratory of Duke University, certainly appeared very remarkable from the reports of the tests which aimed at eliminating all possible sensory cues; Dr. Pratt, however, in his book *Parapsychology: an inside view of ESP*, ends an account of them with the words 'the findings must be regarded with a degree of tentativeness'.

The whole question of a possible extra-sensory perception in animals is awaiting a better experimental approach. It is not, however, simply as a means of communication that I think something akin to telepathy may be biologically important. Let me refer back to the British Association address of 1949 in which I first introduced the idea:

> I mention it [I said] merely as a reminder that perhaps our ideas on evolution may be altered if something akin to telepathy—unconscious no doubt—was found to be a factor in moulding the patterns of behaviour among members of a species. If there was such a non-conscious group behaviour plan, distributed between, and linking, the individuals of the race, we might find ourselves coming back to something like those ideas of subconscious racial memory of Samuel Butler, but on a group rather than an individual basis. . . . If there was such a group habit and behaviour pattern it might operate through organic selection to modify the course of evolution: working through selection acting on the gene complex. If this flight of fancy ever proved to be a fact, it would be a

wedding of the ideas of Darwin and Mendel on the one hand and of Lamarck and Samuel Butler on the other!

I am now not surprised that my audience did not follow me. I had not realized that in the 1940s there was not more than a handful of zoologists who remembered what was meant by the technical term of Organic Selection although I had taken pains to remind them a little earlier in my address:

> A still more important contribution [I said] that field zoology can make to evolutionary theory is to throw more light on the part played by Organic Selection. The gene combinations which are best suited to the *habits* of the animal may tend to survive in preference to those which do not give such full scope to the animal's pattern of behaviour. This idea of Organic Selection, which was put forward independently by Baldwin and Lloyd Morgan at the turn of the century, has been almost forgotten until quite recently. This possible selection of structural variations by habit as opposed to the selection of other variations by the environment may indeed be a factor of importance. It is in effect similar to that postulated by Lamarck but brought about on Darwinian lines. External Natural Selection must of course be important, but if Organic Selection can be shown to be a really significant factor, it may well alter our way of looking at evolution as a whole. The relative importance of the two forms of selection must be the subject both of experiment and of more research into the habits and behaviour of animals in nature.

Because this principle enunciated by Baldwin and Lloyd Morgan had been so largely forgotten by the present generation I devoted two of my recent Gifford Lectures to the subject (Lectures VI and VII) published in *The Living Stream*. Since the idea is essential for my theme—that something like telepathy may be an important biological principle—I will devote a little space to it here. I am a neo-Darwinian in that I believe evolution, on its physical side, is brought about by the action of selection upon the ever-changing chemical genetical material which is handed on from generation to generation; I do not, however, regard this as the whole story. There is another part of the process which I consider to be of equal or even greater importance; this concerns the nature of the selective agents operating in it. Most biologists today think of natural selection as being entirely an external force: either of the physical evironment, selecting those varieties of a species best suited to a particular climate or habitat, or of the animate environment, the action of predators or competitors, eliminating those which are less fleet of foot or less well concealed by colour or protected by other devices and so on. In addition to the selection made by all these factors in the environment, there is another selecting force, and this is not external; it is the action of a *change of habit* developing within a population of animals, which itself becomes a selecting agent. It is true that often the change of habit is brought about by changes in the environment such as a shortage of the normal kind of food, the destruction of typical breeding sites and so on, but *not always*; a change of habit, as we shall see, may also be brought about by the inquisitive exploratory nature of animal life, so different from that of a plant.

Change of habit, as a factor in evolution, was the element stressed by Lamarck. In saying this I must explain that I am not a Lamarckian in the generally accepted sense of the term; I do *not* believe that change of habit can influence evolution through a supposed inheritance of changes in bodily structure brought about directly by a greater use of some organs or a lesser use of others. I certainly feel, however, that Lamarck deserves much more credit than he gets at present, for discerning the importance of the behavioural side of the animal in the working of evolution; I have little doubt that he will get it in time.

The way in which I believe change of habit affects the course of evolution is by a particular form of selection: a selection of the more suitable combinations of the random genetic mutations, but one brought about by the *behaviour* of the animal itself. When I have previously discussed this I have, I think, usually been misunderstood; in trying to avoid this I may perhaps over-labour one or two points. I want to be quite certain of making clear what some people evidently find to be a somewhat subtle distinction. The ever-varying hereditary material, changing through the continual reassortment and recombination of the genes which from time to time undergo random mutations, may be conveniently referred to as the *gene complex*; for brevity I shall use this term in this sense. Now *if a population of animals should change their habits then, sooner or later, variations in the gene complex will turn up in the population to produce small alterations in the animals' structure which will make them more efficient in relation to their new behaviour pattern;* these more efficient individuals will tend to survive rather than the less efficient, and so the composition of the population will gradually change—a *change being due initially to a change in habit.*

Let me give an illustration. If, for example, birds of a particular species, originally feeding on insects from the surface of the bark of trees, found, in a time of shortage, that they could get more by probing into and under the bark, then they might develop a change in habit which could spread through the whole species population, just as we have seen the new habit of opening milk bottles spread (apparently by copying) right through the tit populations of Europe. Now if this new habit became well established, then any members in the population with a gene complex giving a beak slightly better adapted to such probing would have a better chance of survival than those less well equipped. A new shape of beak would be evolved as a result of a change of habit. The same sort of thing will happen in any group of animals; when a mammal turns more to digging or climbing for its food, or diving into water after fish, we shall see a modification of bodily form brought about through a change of behaviour. Members of a terrestrial species do not by mere chance first get webbed feet and then take to the water to use them; a population develops an aquatic habit and then, in subsequent generations, any new variations with more efficient swimming feet will have a better chance of survival. This all seems most obvious, yet I believe the real point of it is missed by most of those who have considered it.

When I have discussed this before, I have thought that I was merely recalling the views which had been put forward by Baldwin in America and Lloyd Morgan in this country at the turn of the century and which, as I have said, they called Organic Selection. Now some people tell me that I am reading too much into Baldwin's views and that what I am saying is really something different. Most people, however, when I discuss it, tell me that I am simply talking about pure Darwinism and that I am just making a fuss about nothing. This is where I disagree and insist that there is a real if subtle difference which is fundamental to all that we are discussing. I will try and explain why.

No one will deny the great importance of selective forces working from the outside on the members of an evolving population of animals. All the wonderful adaptations, involving colour, shape of body, and even behaviour patterns, which together camouflage an animal to resemble some other object, say, a leaf or stick, could have been produced only by changes in the gene complex being selected from outside the individuals, in fact by the action of predators which fail, on the whole, to secure the better likenesses; they could only be produced in this way because camouflage has no significance unless viewed at *some little distance* from the animal concerned. Now equally important, I believe, are these other selective forces originating in the behaviour of the animal itself. Because a change of habit is usually occasioned by changes in the environment, it is generally supposed, I think, that any selection due to such a habit change is one differing only in degree, but not in kind, from the other forms of selection just discussed. This for me is the crux of the whole issue. I realize, of course, that the differential mortality allowing the more efficient beak, for example, to survive is brought about by factors in the external world, but nevertheless the real agent at work is the behaviour pattern within. I believe the case for regarding this 'behavioural selection' (i.e. selection resulting from changes in behaviour) as radically different can be defended.

Before going any farther let me discuss a little more fully the character of the adaptations brought about by such behavioural selection. I think, on the whole, they may be said to be the more fundamental in determining the animal's make-up. Those adaptations produced by external selection are generally more limiting or negative in nature, pruning the organism to fit the physical environment or supplying it with better means of escaping the attacks of predators. The adaptations due to the selection made by the animal's own pattern of behaviour, are surely the ones that distinguish the main diverging lines of any evolving group of animals leading to different roles of life, to forms better adapted and specialized, for example, for running, digging, climbing, swimming, or flying, etc. And this is true not only of the vertebrates. Dr. Sidnie Manton in her beautiful studies of arthropod locomotion and feeding mechanisms stresses again and again that habit must be the overriding factor in the evolution of the different types of bodily form encountered. While we must, of course, give due weight to the pruning and limiting effects of natural selection from outside, these are of less significance compared with those of the creative behavioural selection from within which are for ever giving rise to novelty.

The distinction I am making between the environmental and behavioural selective forces is well illustrated by one of the outstanding differences between the plant and animal kingdoms. The two stems of organic life evidently separated in evolution before the development of behaviour patterns such as we know in animals. The greater part of a plant's structure is mainly, perhaps entirely, the result of external evironmental selective forces, for example, those of the physical environment, of competition with neighbours, and of the browsing of herbivorous animals. Now animals are subject to as many such external forms of selection as are plants, but, in addition, they have this behavioural selection which makes the vital difference. It is instructive to note that the more dramatic adaptations of plants are those of flower structure, and while these are produced by an external selection, it is one brought about by the behavioural patterns of pollinating insects. The elaboration and beauty of flowers are the products of animal behaviour.

Those who are not already familiar with the Baldwin–Lloyd Morgan concept of organic selection are not likely to be convinced of the importance of this extension of the principle, which may be called behavioural selection, just by this very short sketch of the idea; the complete arguments in support of it should be examined in the full account of it which I give in *The Living Stream*. Let me, however, assume that we can now regard this system as a valid part of evolution theory, at any rate among the higher vertebrates, such as birds and mammals, where we can actually see new behaviour patterns spreading through a population as in the recent opening of milk bottles by tits, or in the new habit of taking the seeds of the shrub Daphne by greenfinches which has been observed to have gradually spread across Europe in the present century. Now the possible existence of something akin to human telepathy at a subconscious level among animals is not essential for the working of this behavioural selective system among the members of the higher vertebrate species where, as we have just seen, the spread of tradition by copying does in fact take place; but it would most likely facilitate the process.

The rather unusual human beings with telepathic gifts may be presenting us with chinks in the material carapace of the living world through which we can probe bit by bit to find out a little more of what at present appears a mystery lying behind it. The existence of such an unconscious telepathic communication between members of the same species of animals might at least help in developing and stabilizing common behavioural patterns. The late Mr Whately Carington, who made some very interesting experiments in the transmission of drawings some twenty years ago, put forward such an idea in his book *Telepathy*, published in 1945. After discussing the remarkable web-spinning activities of spiders, he writes: 'I suggest that the instinctive behaviour of this high order or elaborate type may be due to the individual creature concerned (e.g. spider) being linked up into a larger system (or common subconscious if you prefer it) in which all the web-spinning experience of the species is stored up.' Actually there can be little doubt that truly instinctive behaviour is governed by the DNA gene complex—probably built in by organic selection from gradually formed habits of increasing elaboration. These habits, however, before being genetically fixed, might be spread and stabilized in the population by some such telepathy-like means, for we have seen that they must become widespread before they can be so incorporated. Such a hypothesis might help to explain the development of elaborate instincts in invertebrate animals, among which it would be difficult to conceive of new habits spreading by copying and tradition.

This idea, of course, is entire speculation and we must not jump to the conclusion that it is in fact true; I am merely suggesting possible ways in which the existence of something like telepathy, if proved to exist among animals, might modify our views as to the whole process of evolution.

The mechanism whereby new habits do spread through populations of invertebrate animals is indeed an unsolved problem at present and one which must be investigated experimentally. Some of the most beautiful examples of what must surely be the result of a behavioural selection are to be found among invertebrate animals. I have already cited the views of Dr. Sidnie Manton on the importance of habit in relation to the evolution of structure among the arthropods, but I will give another example from the work of Dr. A. J. Alexander[7] in which she makes an analysis of the evolution of different stridulatory mechanisms in

[7] *Proc. Zoolog. Soc. of London*, Vol. 133, 1960, pp. 391–8.

related species of scorpions. One species of the genus *Heterometrus* in a 'threat situation' would adopt an aggressive attitude with the claws (the pedipalps) thrust forward in a clasping attitude, whereas another species, of the genus *Pandinus*, made more defensive movements with the claws pulled in to cover the front of the animal. They both produce a threat sound, a rasping stridulation, by the development of 'keyboard' bristles, but on the *pedipalp* of *Pandinus* and on the *first leg* of *Heterometrus*. They have been developed from the unspecialized bristles in the two areas entirely in relation to the differences in the movements in the two kinds of threat behaviour.

I now want to turn to what is a somewhat more technical, but very important, problem in biology: that of 'homology'. It is a concept which is fundamental to what we are talking about when we speak of evolution and appears on the surface to be both a very simple one and one easy to describe; it is not, however, I believe quite so easy to account for as has usually been thought. What we mean by the term may be readily explained by an example. Let us consider the bones of our arm: in the upper arm is a single bone, the humerus, whereas the forearm has two more or less parallel bones, the radius and ulna, which articulate with the group of wrist bones (carpals) and these in turn with the bones of the hand (the metacarpals and phalanges). We find the same *general* plan in the fore limbs of all the terrestrial vertebrate animals, whether they be amphibians, reptiles, birds, or mammals, but with the different bones varying in relative size and shape to suit the activities of the creatures concerned; some limbs are beautifully adapted for running, others for climbing or digging and some still further modified for flying, as in the wings of birds or bats, or for swimming, as in the paddles of a turtle or the flippers of a whale. In some species two bones, such as the radius and ulna, may be fused together or in some others the number of digits (fingers) may be reduced from the typical five to four, three, two, or even one (in the case of the horse) by the loss of bones. The actual *layout* of the plan, however, is never altered; we never, for example, find *two* bones in the upper arm, always one, the humerus. The humerus bones throughout the whole vertebrate series exhibit what we mean by homology and may be said to be homologous; similarly, each of the other parts of the forelimb, the radius, ulna, carpals, etc., are homologous, and, of course, one whole forelimb is homologous with another. So indeed are all other parts of the vertebrate skeleton, likewise the parts of other organs, for example the lens, the iris and the retina of the eyes of all vertebrates are regarded as homologous. The term homology was invented to contrast with analogy. The wing of a fly and that of a bird are analogous in that they are both used for the same purpose, flying, but they are *not* homologous because they are constructed in entirely different ways; the wing of a bat and the flipper of a whale are homologous, but not analogous because they serve quite different functions.

What do I mean by saying that the concept of homology presents a problem, a puzzle? Surely if all the vertebrate animals are descended from a common ancestor, the explanation of homology must be as simple as could be: each of the homologous organs, in their almost endless variety, will be but modifications of the corresponding organs in the original progenitor of the race. This in one sense is true, *but* modern genetical research has now turned up some facts regarding the mechanism of the process which complicate the issue and make it not quite so simple as it appears on the surface.

Before I go farther it is worth recording that the term homology actually goes back in the history of biology to pre-evolution days; perhaps the simplest definition of it is that given in 1843—sixteen years before the publication of Darwin's *Origin of Species*—by the great anatomist Richard Owen who was always a bitter opponent of the evolution doctrine. He defined a homologue as 'the *same* organ in different animals under every variety of form and function' and contrasted this with an analogue which, he said, is 'a part or organ in one animal which has the same function as *another* part or organ in a different animal'. The early naturalists of this period considered the principle of homology to be an essential element in the Deity's design of the world: a plan upon which He played an almost infinite variety of changes, by lengthening this or shortening that, to give all the marvellous diversity of creation. They thought that God must have invented a series of different general plans, which they called archetypes, for the different main groups of animals; such were the arche-vertebrate, the arche-articulate [8] and the arche-mollusc types—the key ground plans for the vertebrate, the arthropod and molluscan

[8] The arthropods used to be called the Articulata.

animals. When evolution came to be accepted the idea of homology fell at once into place; the supposed archetypes of the plan of creation became the ancient progenitors of the different races: the ancestral vertebrate, the ancestral arthropod, ancestral mollusc, and so on. It all seemed so very simple and obvious.

When I was an undergraduate student just after the First World War, and indeed when I was a professor in the thirties, there seemed to be no problem at all. The same homologous structures must clearly be due to the same hereditary factors handed on generation after generation from the early ancestor with occasional changes by mutation; the wide variety of form seen in different animal groups being due to natural selection acting upon these factors or genes which were handed on, with mutational changes, from the original ancestral form. With the development of experimental genetics, however, the old idea of one factor or gene governing one particular body character has been replaced by that of the gene complex whereby all the genes are interacting to have their united effect upon the various structures. In truth we can no longer say that homologous structures are always due to the same—homologous—genes, however modified by mutation, handed on in the process of descent. Any animal structure we are looking at is produced by the combined effects of a particular gene complex and the influence of the environment in which the animal develops; and we now find that what we have been calling homologous structures are often produced by the action of *quite different* genes.

T. H. Morgan was perhaps the first to demonstrate this surprising fact in 1929;[9] although its importance was not recognized until a number of other examples came to light, particularly those discovered by S. C. Harland,[10] among plants. In the fruit fly *Drosophila* there is a particular gene which governs the formation of the eyes and there is an allelomorph (a mutant alternative) of this gene which in the homozygous state (i.e. in the twofold condition) produces an eyeless condition. Now Morgan showed that, if a pure homozygous eyeless stock is inbred, the other genes in the gene complex, by reassortment, may come to be recombined in such a way that they will deputize for the missing normal eye-forming allelomorph, and, lo and behold, flies appear in the

[9] T. H. Morgan, *Publ. Carnegie Inst. Wash.*, Vol. 399, 1929, pp. 139–68.
[10] S. C. Harland, *Biol. Rev.*, Vol. II, 1936, p. 83.

'eyeless' stock with eyes as good as ever! These eyes must surely be regarded as homologous with the eyes of normal flies, yet their production is not controlled by the same genes. Homologous structures need not be produced under the influence of homologous genes. Rather similar results were obtained by Gordon and Sang[11] with a stock of *Drosophila* lacking antennae; in a culture of such antennaless flies, they were able by inbreeding, and so reshuffling the gene complex, to produce flies which had either only one antenna or the normal pair. Several other similar effects in the little fly *Drosophila* were shown by Mohr.[12] Again, as Fisher showed,[13] structural characters controlled by identical genes need not be homologous; in wild-type poultry he showed that a particular gene acted as a dominant when producing a crest of feathers but as a recessive in respect to a condition called cerebral hernia in which the frontal and parietal bones of the skull fail to close so that the brain bulges out through the gap. In certain breeds of poultry other genes in the complex completely suppress the hernia effect but allow the crest of feathers to appear. Then, indeed, all Waddington's experimental genetic assimilation effects such as cross-veinless and others show us how apparently identical characters may be brought about by quite different assortments of genes.

The concept of homology in terms of similar genes handed on from a common ancestor has broken down. Perhaps homologous structures are always formed from the same corresponding set of *cells* in development? No, this also fails; the lenses of the vertebrate eyes must surely be regarded as homologous, yet in experiments on frogs and newts they may be formed from epidermal cells at all sorts of places on the body surface if part of the eye (the optic cup) is grafted in below the skin. The optic cup is then said to act as an 'organizer'. It was thought then that homologous structures might be due to the handing on from ancestors or similar 'organizers'; this hypothesis, however, also collapses. For instance, in one species of frog (*Rana fusca*) the lens of the eye can only be induced by the presence of the optic cup; in another species (*Rana esculenta*) while it can be induced by the optic cup, it is also formed in its proper place if the optic cup is removed—formed apparently in relation to the developing whole animal.

[11] C. Gordon, and J. H. Sang, *Proc. Roy. Soc. B.*, Vol. 130, 1941, pp. 151–84.
[12] C. Mohr, *Z. f. ind. Abst.*, Vol. 50, 1929, pp. 113–200.
[13] R. A. Fisher, *Phil. Trans. Roy. Soc. B.*, Vol. 225, 1935, pp. 195–236.

This subject has been well discussed by Sir Gavin de Beer,[14] but he does not, I think, experience my difficulties.

For the present we appear to be forced into the position of saying that the only explanation of homology that the latest generally accepted views on evolution can offer is that *selection by the environment* is governing the maintenance of all the internal spatial relationships of the animal; i.e. all the multitude of homologous parts which make up complex creatures such as, say, a hedgehog, a chaffinch, or a frog. We must recognize that within relatively short periods of time there is a good interchange of genes (gene flow as it is called) throughout the range of an interbreeding population and this helps to keep the race comparatively uniform; is it not, however, stretching the concept of *external* selection a bit far to suppose that it alone, by controlling the effects of an ever-changing gene complex, is maintaining the stability of structure in a species over vast areas of *different types* of country—and over long periods of time? Can the whole complicated *internal* structure of our chaffinch, for instance, really be maintained—or rather slowly evolved—entirely under the influence of its multifarous *external* surroundings and nothing else? I could understand natural selection by the environment controlling the evolution of the whole intricate organ system if there were, associated with the homologous structures, some actual homologous units which varied and were handed on to be selected. But no, the homologous structures now appear to be governed by the *effects* of a whole multitude of units which are continually being reassorted. According to modern mechanistic biology the DNA 'plan' for the intricate homologous 'machinery'—for instance, the vertebrate, the arthropod or the molluscan plan—would seem to have been laid down by the selective forces of the variable environment outside. I am perfectly prepared to accept the proposition that the DNA genetic code is handing on from generation to generation the specification for the plan of development of the animal body as determined by an act of selection of one sort or another; I *am* doubting, however, whether the plan itself is entirely the product of the environment. To my way of thinking, and remembering the great variety of environments which a single species may encounter and the variety of different kinds of

[14] G. R. de Beer, 'Embryology and Evolution', in *Evolution*, ed. de Beer, Oxford, 1938.

animals which may live in the same habitat, such a conclusion seems almost a *reductio ad absurdum*. I may be unduly sceptical but I cannot help wondering if there is not something else concerned with the evolution process that we do not yet understand.

Among the experimental demonstrations of telepathy which have impressed me most are those whereby drawings of various designs were transmitted from one person to another under most carefully controlled test conditions, such as those of Guthrie and Birchall in which Sir Oliver Lodge took part,[15] or the experiments in which Gilbert Murray received telepathically from members of his family impressions of quite elaborate human situations selected from passages in books and other sources.[16] If such impressions of design, form, and experience can occasionally be transmitted by telepathy from one human individual to another, might it not be possible for there to be in the animal kingdom as a whole not only such a telepathic spread of habit changes as I have already suggested, but a general *subconscious* sharing of a form and behaviour pattern—a sort of psychic 'blue print'—shared between members of a species? It is important to remember that in the concept of the individual mind we are faced with a mystery no less remarkable. The mind cannot be anchored to this or that group of cells that make up the brain. The community of cells making up the body has a mind beyond the individual cells—the 'impression' coming from one part of the brain receiving sensory impulses from one eye and that from another part of the brain from the other eye are merged together in the mind, not in some particular cells as far as we know. The relation of the mind and body seems at present to be as inconceivable as my speculation.

If there were such a 'psychic' plan it would be something like the subconscious racial memory of Samuel Butler, but it would be a racial experience of habit, form, and development, open subconsciously to all members of the species, as in Whately Carington's group mind, or as in Jung's shared unconscious, if I understand him aright. I always feel that Butler's idea, although it broke down, was a brilliant one: a remarkable evolutionary conception developed by an amateur genius;[17] he thought that the

---

[15] *Proc. S.P.R.*, Vol. 2, 1884, pp. 24–42.
[16] *ibid.*, Vol. 29, 1916, pp. 64–110, and Vol. 34, 1924, pp. 212–74.
[17] *Life and Habit* (1878), *Unconscious Memory* (1880), and *Luck or Cunning*

egg carried on a subconscious memory from the parent of how to develop, of how to manipulate its protoplasm, of how to carry out its instinctive habits.

In the fanciful idea I am suggesting—and at present it is nothing more—a psychic pool of experience would be shared subconsciously by all members of a species by some method akin to what we are witnessing in telepathy. Individual lives, animal 'minds', would come and go—but the psychic stream of a shared behaviour pattern in the living population would flow on in parallel to the flow of the physical DNA material. There would be two streams of information—the DNA code supplying the varying physical form of the organic stream to be acted on by selection—and the psychic stream of shared experience—the subconscious species 'blueprint'—which *together with the environment*, would select those members in the population better able to carry on the race. External conditions being equal, those animals with gene complexes which allow a better incarnation of the species plan would tend to survive rather than others whose gene complexes produced less satisfactory versions. Such an internal conserving selective element might explain the secret of homology in face of an ever-changing gene complex. And then again it might also be supposed that the 'racial plan', linking all the members of the race, might gradually change as the character of the population became modified *both* by the changing environment's external selection and by the development of new behavioural patterns due to the exploring, exploitive, nature of animal life.

I do not really expect that my suggestion will be found to be true; I use this somewhat extravagant speculation as an illustration of the kind of way in which the psychic and physical sides of life may be found to interact. Of course the truth will be found to be very different from, but no less remarkable than, what we can imagine at the present time; I feel certain that what we are now recording as extra-sensory perception must have a much greater biological significance than a mere occasional telepathic transmission. That is why I regard psychical research as so important for biology; I believe it will open the door to a deeper understanding of life than that which can be given by the physico-chemical approach alone.

(1889). There were other mnemonic theories of course, such as those of Hering and Semon, but they lacked the vision of Butler's views.

VII

THE NOTION OF 'PRECOGNITION'

C. D. Broad

I will begin by offering a few selected cases which, between them, illustrate some of the various forms of the phenomena which I intend to cover by the phrase 'ostensible precognition'. They are well enough attested to be worthy of serious consideration, but I am not concerned here primarily with questions of evidence. The cases may be divided into 'Experimental' and 'Sporadic'. I will begin with the former.

EXAMPLES

Experimental

As my one example I will take Dr. Soal's experiments on card-guessing, with Mr. Shackleton as guesser.[1] The essential points are these. Shackleton knew that the 'telepathic agent', on each successive occasion in the course of an experiment, would be looking at a picture of one or another of a certain five different animals, the names of which were already known to him. On receiving a signal he would record in writing the initial letter of the name of the animal which he guessed the agent to be looking at then. The order in which the various picture-cards was presented to the agent for inspection was random, and the interval between successive presentations was about 2·5 seconds. The experiments were conducted in runs of twenty-five, with an interval between each run.

Now the initial letter written down by the guesser on the nth

[1] For details, see S. G. Soal and F. Bateman, *Modern Experiments in Telepathy* (London: Faber, 1954).

occasion may happen to be the initial letter of the symbol presented for inspection to the agent on the nth occasion. If so, we may say that the guesser scored a '*direct* hit'. On the other hand, it may happen that the nearest occasion on which the card looked at by the agent has on it the animal whose name begins with the letter written down by the guesser on the nth occasion, is the $(n-p)$th, or, alternatively, that it is the $(n+q)$th. On the first alternative we should say that the guesser on the nth occasion had scored a '$-q$-back-hit'; while, on the second alternative, we should say that he had scored a '$+q$-fore-hit'. It is obvious that it is $+q$-fore-hits, and they alone, which are relevant in these experiments to the question of precognition.

Dr. Soal deliberately confined his attention to the following five alternatives, viz. -2 and -1 back-hits, direct hits, and $+1$ and $+2$ fore-hits. The question, with regard to each of these alternatives is this: 'Assuming that the order in which the sequence of pictures was presented to the agent was random, did the number of hits of a given kind (e.g. direct hits, or $+1$ fore-hits) differ significantly from the number most probable on the hypothesis of mere chance-coincidence between picture guessed and picture presented? And, if it did, what were the odds against at least as great a deviation occurring, on the hypothesis of chance-coincidence, within the number of guesses actually made?'

I pass now to the results, so far as they concern $+1$ fore-hits. Of the thirteen different persons tried as agents there were only three with whom Shackleton scored significantly large deviations in any of the five positions. These were two women, R.E. and G.A., and one man, J.A. I shall here confine my attention to the results obtained with the two women. With both of them there were no significant deviations except in regard to $+1$ fore-hits.

Taking together all the trials done with R.E. and all those done with G.A. as agent, we have in all 5,799 guesses which *could* have resulted in $+1$ fore-hits. Of these 1,679 *in fact* did so. On the hypothesis of chance-coincidence the most probable number of such hits is 1,160 to the nearest integer. The excess of $+1$ fore-hits over the number that is most probable on the hypothesis of chance-coincidence is therefore 519. In calculting the odds against so large a deviation occurring by chance-coincidence, we must, of course, allow for the fact that $+1$ fore-hits is only one of the five possibilities under consideration. But, even when we allow for this,

THE NOTION OF 'PRECOGNITION'

we find that the odds, on the hypothesis of chance-coincidence, against getting so great a deviation in one or another of the five positions under consideration, are $2 \cdot 4 \times 10^{63}$ to 1. They are about the same as the odds against throwing not less than 82 sixes in succession with a fair die, when one starts throwing and continues until a non-six turns up.

We may sum up these results as follows. There is a not very strong, but extremely persistent, positive association between the nature of Mr. Shackleton's guess on any occasion and the specific nature of the perceptual experience which the agent *will first begin to have a few seconds after the guess has been recorded*, viz. when she shall be presented with a picture *on the next ensuing occasion*. This positive association is not very strong. It is an actual percentage of 28·95 per cent $+1$ fore-hits, against the 20 per cent which is most probable on the hypothesis of chance-coincidence. But it is so persistent that the odds against so strong an association persisting as a mere chance-coincidence, in a run of guesses so long as that actually made, are colossal. They are such as to rule out that hypothesis completely. It is for this reason that we say that the results of these experiments seem *prima facie* to establish the occurrence of 'precognition' on the part of Mr. Shackleton.

Before leaving this example, I would like to emphasize the following facts about it:

(i) We have spoken of Mr. Shackleton's activities as 'guessing'. That might suggest that on each occasion he made a special effort to envisage mentally (e.g. by means of an appropriate visual or auditory image) the picture on the card at which the agent was then, or would immediately afterwards be, looking. Any such suggestion would be quite misleading. There is no evidence that the movements of his hand, in writing down now this and now that initial letter, were made in response to any relevant imagery, or were in any way premeditated. And, when we consider the very rapid rate at which the calls were made and the letters written down, it seems plain that what we have called 'guessing' is no more than the almost mechanical writing down of one or another of the five initial letters, without having any conscious reason on any occasion for writing down any particular one.

(ii) What is consciously present to the guesser's mind throughout any such experiment is a knowledge of the general experimental set-up, and a general intention to try to respond on each occasion with the appropriate initial letter, whatever that may be. It should be noted, however, that what Mr. Shackleton in fact accomplished was *not* what he was consciously trying to do. What he was consciously trying to do was to write down the initial letter of the name of the animal depicted on the card at which the agent was *then* looking. What he in fact accomplished, to an extent far beyond all question of chance-coincidence, was to write down the initial letter of the name of the animal depicted on the card at which the agent *would be looking on the next ensuing occasion*. It would therefore be more accurate, and less question-begging, to describe his performance as '*pre-presentative verbal response*' than to call it 'precognition'.

(iii) In experiments conducted under similar conditions with the agent Mr. J. A., Mr. Shackleton scored a highly significant excess over chance-expectation, not only on $+1$ fore-hits, but also on -1 back-hits. Roughly the same proportion of his successes was of each of these two kinds. He was thus displaying, in the same run of guesses and to about the same degree, both pre-presentative and retro-presentative verbal response. But he appears to have been quite unaware that he was doing so, or indeed of any introspectable difference between the responses which were in fact of the one kind and those which were in fact of the other.

(iv) Sometimes Mr. Shackleton had a 'hunch' that he was being highly successful, while at other times he felt that he was accomplishing nothing but random responses. It was found that there was no significant correlation, positive or negative, between these subjective impressions and the actual degree of success or failure of a run of guesses.

I have mentioned these four points in order to bring out the extremely behaviouristic character of Mr. Shackleton's performance. This might easily be overlooked, if we were to describe what he accomplished as 'precognition', without further comment.

2. *Sporadic cases*

The case just described is experimental. It may also be called 'quotitative'. What I mean is this. There is nothing at all remarkable in *any particular* guess turning out to be a $+1$ fore-hit; what is re-

THE NOTION OF 'PRECOGNITION'

markable is the excess of the *aggregate number* of such guesses over the number most probable on the hypothesis of chance-coincidence. I pass now to cases which are *sporadic* and *non-quotitative*. Here we have a single experience (or, in a few cases, several experiences occurring at irregular intervals in the same person), e.g. a dream or a waking hallucination or a felt impulse to do or to avoid doing a specific action. This experience is highly detailed and peculiar. It is followed, after a fairly short interval, by a single incident or state of affairs, which is also highly peculiar, and which seems to correspond in detail with the earlier experience in a way and to a degree that rules out chance-coincidence.

The reader who wishes to make himself familiar with a collection of cases, well arranged and critically discussed, cannot do better than consult H. F. Saltmarsh's admirable 'Report on Cases of Apparent Precognition' in the *Proceedings of the S.P.R.*[2] He should also read J. Fraser Nicol's 'Apparent Spontaneous Precognition' in *International Journal of Parapsychology*.[3] Here I will merely cite two cases, to indicate the kind of incident which we have to consider.

2.1. *Case of Mrs. Verrall and Mr. Marsh.*[4] Mrs. Verrall was engaged fairly regularly in automatic writing from March 1901 till 1903. On 11 December 1901 her hand wrote: 'Frost and a candle in dim light. Marmontel. He was reading on a sofa or on a bed—there was only a candle's light. She will surely remember this. The book was lent—not his own. He talked about it.'

Marmontel was not an author with whom Mrs. Verrall was acquainted except by name, and she could make nothing of this. Nor could Mrs. Sidgwick, whom she consulted by letter soon afterwards.

On 17 December, six days later, Mrs. Verrall was disturbed all day by a strong impulse to write. She resisted until 6.30 p.m., when she sat down and allowed her hand to scribble. The script ran as follows: 'Marmontel is right. It was a French book; a memoir, I think. "Passy" may help—"Souvenirs de Passy" or "Fleury". "Marmontel" was not on the cover. The book was bound and was lent. Two volumes in old-fashioned binding and

[2] *Proc. S.P.R.*, Vol. XLII, 1934.
[3] In *Int. J. Parapsychol.*, Vol. III, 1961, p. 26.
[4] See Mrs. A. W. Verrall, 'On a Series of Automatic Writings'. *Proc. S.P.R.*, Vol. XX, 1906, pp. 331 ff.

print.... It is an attempt to make someone remember an incident.'

There the matter rested until 1 March 1902, on which date Mr. Marsh, a friend of the Verralls, came to spend the weekend with them. At dinner he happened to mention that he had been reading *Marmontel*. Mrs. Verrall pricked up her ears. In answer to a question by her, Mr. Marsh said that it was Marmontel's *Memoirs*, and not his 'Moral Tales', which he had been reading. He had *borrowed* the book from the London Library, and had taken the first volume with him to Paris, where he had read it on the evenings of 20 and 21 February, i.e. some ten weeks after the first, and some nine weeks after the second, reference to Marmontel in Mrs. Verrall's script. On both occasions he read it *by the light of a candle*; on the 20th he was reading it *in bed*, and on the 21st lying *on two chairs*. He had *talked about* the book to the friends with whom he was staying in Paris. The weather was *cold*, but *not* frosty. Asked by Mrs. Verrall whether the references to 'Passy' and to 'Fleury' were relevant, Mr. Marsh deferred his answer until he should have returned to his home in London and have had an opportunity to look up the book. On 24 March he wrote from London to say that on 21 February, while in Paris, he had read, lying on two chairs, a chapter in the first volume of Marmontel's *Memoirs*, describing the finding at *Passy* of a panel, connected with a story in which *Fleury* plays an important part. Mrs. Verrall subsequently ascertained that these two names do not occur together anywhere in Marmontel's *Memoirs* except in the passage in question. Mr. Marsh states that he had not read anything by candlelight, lying on two chairs, for months before the occasion when he did so in Paris on 21 February.

There are two remarks worth making on this case. (i) Although the two passages in Mrs. Verrall's script seem plainly, on retrospect, to be correlated with what Mr. Marsh would be doing in Paris some two months later, they did *not* present themselves as referring to the future. On the contrary, they presented themselves in the script as referring to something that *had already happened*, which it was hoped to recall to the memory of someone (whether of Mrs. Verrall herself or of someone else, is not clear). (ii) There are certain minor discrepancies in detail. The weather was *not frosty*, though it was cold. The name 'Marmontel' *was* on the binding. Mr. Marsh had *only one* volume of the book with him in Paris,

THE NOTION OF 'PRECOGNITION'

though he did read a second volume on his return to London. There were, in fact, three volumes in all in that edition.

2.2. *Case of Mrs. C and the escaped monkey.*[5] On 29 February 1888 Mrs. C of Holland Road, Kensington, wrote to Myers, describing the following incident which had happened to her in 1867.

She had always had a horror of monkeys, and one night she dreamed that she was persistently followed by one, which terrified her, and from which she tried in vain to escape. Next morning she told her dream to her husband and other members of her family. He advised her to take a short walk with the children in order to get rid of the impression. Quite contrary to her usual custom, she did so, taking with her her children unaccompanied by their nurse. In a narrow lane (Holland Lane) she passed Argyll Lodge, and there saw, on the roof of the coach-house, the monkey of her dream. The monkey began to follow her, he on the top of the wall, and she and the children in the lane below; and she experienced the same terror as in her dream, fearing every moment that it would jump down on them. It did not do so, and eventually she reached home safely, in a great state of agitation. Shortly afterwards she sent someone round to Argyll Lodge to inquire, and was informed that on that morning a very rare and valuable monkey, belonging to the Duchess of Argyll, had got loose. Myers received letters from Mr. C and from Mrs. C's nurse, confirming Mrs. C's story from their own recollections. He was also informed by the Marquis of Lorne that a monkey was in fact kept in the stables of Argyll Lodge at the time in question.

On this case the following comments may be made. (i) No doubt, the psycho-analyst might have his views as to the probable causes of Mrs. C's horror of monkeys. But what has to be explained is the concurrence of that particular dream with that particular incident next morning involving a monkey. (ii) A curious feature is that the occurrence of the dream was a necessary, though not a sufficient, condition of the subsequent events which it prefigured. For it is almost certain that Mrs. C would not have taken a walk that morning, if she had not had, and reported to her husband, the disturbing dream of the night before.

[5] F. W. H. Myers, 'The Subliminal Self', *Proc. S.P.R.*, Vol. XI, 1895, pp. 488-9.

SOME COMMENTS ON THE EVIDENCE ADDUCED FOR PRECOGNITION

The few cases which I have quoted above are intended merely as illustrations. The remarks which follow will be based, not only or mainly on them, but on the varied collection of cases presented in, e.g. Saltmarsh's 'Report', referred to above. We can conveniently consider them under the following three headings, viz. (1) the nature of the allegedly precognitive event, (2) the nature of the allegedly precognized event or state of affairs, and (3) the nature of the correspondence between the two which makes it plausible to hold that the former is a precognition of the latter.

1. *Nature of the allegedly precognitive event*

(i) The allegedly precognitive event may be of many different kinds. In the first place, it need not be an *experience* at all. It may be, as it was in Dr. Soal's experiments with Mr. Shackleton, simply a bit of bodily behaviour, not consciously guided by any relevant thought or image or perception or hallucinatory *quasi*-perception. It is called 'precognitive' only in so far as it turns out to have been such as the individual *would have* performed in the circumstances, *if* he had been aware of certain facts about the future and had been guided by that knowledge. It would be more accurate to describe it as a bit of 'proleptic behaviour' than as a precognition.

Even when the allegedly precognitive event is an actual experience, as it is in most of the sporadic cases, it may be of very different kinds, ranging from a more or less vague emotionally toned impression, through a fairly definite felt impulse to do or to avoid doing some specific action, through imagery (auditory or visual) recognized as such, through dreams (symbolic or imitative), up to full-blown waking hallucinatory *quasi*-perceptions.

Mr. Saltmarsh, in his 'Report', considered all the accounts of alleged cases of precognition received by the S.P.R. in the first fifty years of its existence. There were 349 in all. Of these he rejected 68 altogether, on one ground or another, and was thus left with 281 cases which seemed *prima facie* worth consideration as instances of precognition. He subdivided these into (*a*) 'good', and (*b*) 'ordinary'. In order to count as 'good' the experience reported had to be particularly definite and detailed, and the evi-

THE NOTION OF 'PRECOGNITION'

dence for its occurrence and for that of the relevant later event or state of affairs had to be more than ordinarily satisfactory. Of the 281 cases worth *prima facie* consideration 134, i.e. 47·7 per cent, reached the standard of 'good', as judged by Saltmarsh.

Now he classified the kinds of experience involved in the various cases under six headings, viz. Dreams, Borderline States, Impressions, Waking Hallucinations, Mediumistic Utterances, and Crystal Visions. For our purpose we can take together the dreams and the comparatively few cases where the experience occurred on the borderline between sleep and full waking consciousness. Together they numbered 123, and of them 80 were 'good'.

We may present the relevant statistical facts in the following contingency-table. In it the figure in brackets in each cell gives the most probable number of the 281 cases which would fall into that cell, on the hypothesis of *complete contingency* between the property of being a dream or a borderline experience, on the one hand, and that of counting as a 'good' case of precognition, on the other:

	Dreams or Borderline States	Other Kinds of Experience	
Good	80 (59)	54 (75)	134
Ordinary	43 (64)	104 (83)	147
	123	158	281

$X^2 = 25·45$ (with one degree of freedom) . $p = 6 \times 10^{-7}$

If we calculate from the above table the value of the coefficient of association between being a 'good' case of precognition and being a dream or borderline experience,[6] we find that it is $+0·564$. And the high value of X^2 makes it incredible that so high a degree of association should arise by chance between two characteristics which are in fact quite contingent to each other. So it is plain that there is, among those reported cases of precognition which are *prima facie* worth serious consideration, a strong and highly significant association between being 'good' and being a dream or borderline experience.

I do not think that there is anything surprising in this. For a case does not count as 'good' unless the experience reported was definite and detailed and was such as *could* be closely and

[6] See Yule and Kendall, *Introduction to the Theory of Statistics* (London: Griffin 1958).

unmistakably imitated by a subsequent event or state of affairs in the external world. Now outstanding dreams, of the kind which are likely to be remembered, and to get reported or acted upon, much more often fulfil those conditions than do most of the other kinds of experience considered by Saltmarsh.

I am inclined to think that the fact that so large a proportion of the best evidence for precognition, as regards sporadic cases, relates to dream-experiences, is a serious weakness. The mere fact that dreams are very common, and that most of them are never reported or acted upon, does not particularly matter. For we are concerned here only with those which are sufficiently striking and detailed to be noted and recorded or acted upon very soon after their occurrence. Now these are certainly *not* very common. The weakness is this. One cannot help suspecting that, among such dreams, those which were fulfilled tended, for that very reason, to be submitted to the S.P.R. and similar institutions, while those which were not, tended, for that very reason to remain unpublished. Each of the cases is an instance of an antecedently very improbable concordance between an experience and a later event. If we appeal to the *number* of such cases as a reason for thinking that those concordances are not mere chance-coincidences, we need to be sure that the cases have not been automatically *selected* for attention just *because* they involved such a concordance.

(ii) Whatever may be the nature of the allegedly precognitive event, it does not usually carry with it, for the person in whom it occurs, any *explicit reference to the future*. That it was concerned with something still in the future is generally suspected only later, and often only after a certain event or state of affairs has occurred and has been noted and compared with it.

We have seen, e.g. that Mr. Shackleton took himself to be responding on each occasion to the picture at which the agent was *then* looking. It was only afterwards that it became known that a most improbably large percentage of his responses had in fact corresponded to the picture on which the agent would be focusing on the next ensuing occasion.

When the allegedly precognitive event takes the form of a felt impulse to do or to avoid doing so-and-so, or of an inner voice admonishing or warning one, it does of course explicitly refer to the more or less immediate future. But most other kinds of allegedly precognitive experience, and in particular dreams and

THE NOTION OF 'PRECOGNITION'

waking hallucinations, have no such reference. When one dreams or has a waking hallucination, the scene, the actors, and the incidents are almost always given to one as *present*. On returning to normal waking consciousness one may, in some cases, find cause immediately to refer the recent experience to the future. There is, e.g. a strong tradition that striking dreams about disasters to persons forebode such disasters. If, then, one has had and remembers such a dream, and can find nothing in one's previous experience to account for it, one is inclined to suspect that it refers to a still future event. Again, the circumstances may be such that a certain kind of dream had in those circumstances will inevitably be taken, on waking, to refer to a certain impending future event. If, e.g. the night before the Derby is to be run one dreams of a horse-race and of a certain horse coming in first, one can hardly fail to refer one's dream to what will happen at Epsom next day. But in a great many cases there is nothing to make anyone suspect that such an experience was precognitive until the event occurred which seems, on reflection, to have fulfilled it.

I will call any experience which turns out to have been *prima facie* precognitive a *pro-referential* experience. If an experience, at the time when it occurred, appeared to the experient to refer to the future, I will call it a *pro-spective* experience. Such an experience may turn out to have been pro-referential, or it may turn out that future developments fail to correspond to it or positively conflict with it. If it should prove to have been pro-referential, we may call it *veridical*; if not, *delusive*. A pro-referential experience which was *not* pro-spective may be called *unwittingly pro-referential*.

(iii) It is illuminating to compare the distinctions which we have just drawn, with those which have to be drawn in the case of *post*-cognitive experiences. Take, e.g. the ordinary visual images which from time to time flit before one's mind's eye. Some of them present themselves as referring to some past event, or to some thing, person, or scene witnessed in the past. These may be called *retro-spective*, and may be compared with pro-spective experiences. Like the latter, they may be either veridical or delusive or contain a mixture of veridical and delusive features. Many of our images, however, do not present themselves as referring to the past; and yet it may be possible, in the case of some of them, to show that they almost certainly do correspond to certain past events, or to certain persons, things, or scenes, witnessed in the past. Such images may be called *unwittingly retro-referential*, and may be compared with unwittingly pro-referential experiences.

It is worth remarking that, if a person should have frequently had experiences of a certain kind, e.g. dreams about horse-races, which had proved to be pro-referential, he would come to suspect that *any* such experience in his case would be likely to be pro-referential. He would do this on ordinary inductive grounds. And that conviction might eventually tend, through ordinary association, to give to such experiences a pro-spective tinge which had been lacking formerly.

2. *Nature of the allegedly precognized event or state of affairs*

In considering the nature of the allegedly precognized event or state of affairs. it will be convenient to begin by developing a little further the above analogy between pro-referential and retro-referential experiences. In ordinary retrospection of events the event retrospected, whether it actually happened or not, is always presented as an event in the past history of the *retrospecting subject himself*. If what is retrospected be a person, or a thing, or a scene, or a state of affairs, the latter is always presented as something which the retrospecting subject *himself* perceived, or was concerned in, or heard tell of. We may sum this up by saying that the objects of normal retrospection are *autobiographically restricted*. It is also true that in all normal cases where an experience turns out to have been retro-referential, *without* being retrospective, the past event or state of affairs which corresponds to it is autobiographically restricted.

This limitation appears to most people so nearly self-evidently necessary that any apparent exception to it is regarded as highly paradoxical and is treated as *paranormal*. Examples of such alleged retro-referential experiences, *not* autobiographically restricted as to their objects, are those in which a person has images or hallucinatory *quasi*-perceptions which are found to correspond to historical events, persons, scenes, buildings, etc., which had ceased to exist long before the birth of his present body. (A well-known claim to retro-referential experiences which were *not* autobiographically restricted is the theme of the famous book *An Adventure* by Miss Moberly and Miss Jourdain.) So paradoxical do such claims appear that those who accept them often seek to bring such retro-cogni-

THE NOTION OF 'PRECOGNITION'

tive experiences within the autobiographical limitation, by supposing that the subject existed before the birth of his present body, and witnessed during a previous earthly life the events and scenes which he now retro-cognizes.

Now I think it is true that, in a very large proportion of well-attested cases of ostensible precognition, the future event or state of affairs, which is found to correspond to the experience, is an event in the subject's own later history or is a state of affairs which he will himself witness and perhaps take part in. Cases which do not immediately fall under this heading can often, with a little ingenuity, be brought under it by supposing that what the subject really precognizes is what he will experience or witness when he shall *hear or read of* the future event which corresponds to his precognitive experience. That supposition becomes highly plausible, if it should turn out that the ostensible precognition, though correct in the main, was mistaken in certain details; and if it should be found that the subject never came to witness the events in question himself, but did come to read an account of them which was incorrect in just those details.[7]

But it is certain that there are many alleged cases of so-called 'prophecy' in which the fulfilling event did not happen until long after the prophet's death. Nostradamus, for example, died in 1566. Suppose, for the sake of argument, that the verse in which he seems to be foretelling the flight of Louis XVI to Varennes, and the king's capture there, really was written by him, and that the rather remarkable concordance between the words and the facts is not a mere chance-coincidence. Then the fulfilling event happened some 225 years after the prophet's death.

If we wanted, by hook or by crook, to bring this within the autobiographical restriction, we should have to suppose that the Nostradamus of 1503–66 was re-incarnated, and was again alive in the flesh in 1791. We might then suppose that what the sixteenth-century Nostradamus precognized was not the forthcoming events at Varennes themselves, but was what his eighteenth-century reincarnation would experience when he should read about them in the newspapers.

Now I do not think that anyone would be tempted to make such a supposition in the case of *prima facie* heterobiographical *pre-*

[7] For an example, see J. W. Dunne's *An Experiment with Time* (London: Faber, 1939) 3rd ed, pp. 46 ff.

cognition, while I know that one *is* somewhat tempted to make a parallel supposition in the case of *prima facie* heterobiographical *retro*-cognition. The reason for this difference is, I think, this. *Pre*cognition seems to involve one and the same fundamental *a priori* difficulty, viz. a causal influence of the as yet future on the present, whether it be autobiographically restricted or not. *Retro*-cognition, on the other hand, does not involve any such difficulty of principle. In its autobiographically restricted form it is perfectly familiar to us in the case of ordinary memory, and we account for it causally in terms of 'traces' left on the mind or the brain by past experiences and persisting thereafter up to the present. It is only when it seems *prima facie* to overstep the autobiographical restriction that we feel a difficulty, and we can obviate that by making it really autobiographical in terms of the hypothesis of a previous life in the flesh. Since the fundamental difficulty in admitting *any* kind of *pre*cognition is not in the least lightened by the hypothesis of a future life in the flesh, there is no motive here for making that hypothesis.

3. *Nature of the correspondence between precognized event and precognitive event*

The kind of correspondence between a later event or state of affairs and an earlier experience or bit of human behaviour, which makes one inclined to say that the earlier is a precognition of the later, naturally varies from one class of case to another.

In such experiments as Dr. Soal's the guesser writes down rapidly and almost mechanically on each occasion the initial letter of the name of one or other of a small number of alternatives, which he knows to be mutually exclusive and collectively exhaustive. A hit consists in the letter which he writes down on any occasion being the initial of the name of the picture which the gent *is* focusing on that occasion, *has* focused on a certain earlier occasion, or *will* focus on a certain later occasion. It is only the last of these alternatives that is relevant to precognition. And it is not any *one* such fore-hit that is relevant. Correspondence here consists in the actual proportion of hits at a certain assigned distance ahead, in a long run of guesses, exceeding the proportion most probable on the hypothesis of chance-coincidence to a degree which is highly improbable on that hypothesis.

THE NOTION OF 'PRECOGNITION'

Passing to cases in which the pro-referential event is an actual experience, we may begin with *premonitions*. Here the experience takes the form of a felt impulse, or inner admonition, to take a certain course of action which had not been intended, or to abstain from one which had been intended. Whether this be obeyed or not, there is correspondence provided that the relevant course of events develops in such a way that any normal person in the subject's position would, if he had foreseen it, have wished to behave as the impulse or the admonition had directed.

In cases where the pro-referential experience is an image or a dream or a waking hallucinatory *quasi*-perception there are two possible kinds of correspondence, though both may be mingled in any proportion in a single case. One alternative is that the correspondence may take the form of *resemblance*. The dream or the waking hallucination may be a *quasi*-perception as of so-and-so doing or suffering such-and-such in certain surroundings. And what corresponds may be an actual occurrence at a future date of just such actions by just those agents in just such a scene. The other alternative is that the relationship is *symbolical*. If so, it might be one of a fairly universal kind, as where a dream of a hearse in front of so-and-so's house would symbolize for anyone his death; or it might be peculiar to the experient, depending on his special past experiences and the associations among them. When the correspondence takes the form of *resemblance* we may call the experience *quasi-pre-perceptive*; when the correspondence takes the form of *symbolization* we may call the experience *prefigurative*.

Even when the pro-referential experience takes the form of a *quasi*-preperception of a certain event or state of affairs, which is afterwards realized, it will seldom correspond accurately in all its details. It will generally be supplemented, and often distorted, by features due to the experient's past experiences and acquired associations, his present situation and interests, and so on. When the pro-referential experience is wholly or mainly prefigurative there is, of course, much more room for arbitrary and subjective factors in the correspondence between it and the later state of affairs which it is alleged to have prefigured symbolically.

DEFINITION OF 'X WAS A PRECOGNITION OF Y'

We are now in a position to work towards a definition. It will be more convenient to try to define the phrase 'X was a precognition of Y' than to try to define the word 'precognition'. In reference to what follows I would refer the reader to Professor Mundle's excellent paper 'The Experimental Evidence for P.K. and Precognition';[8] and also to a paper by myself in the *S.P.R. Journal*,[9] in which I discuss an earlier contribution by Mr. W. G. Roll, entitled 'The Problem of Precognition' and comments on it by Professor Mundle and others.

We may begin by dividing the conditions, which are severally necessary and collectively sufficient to make it proper to say that 'X was a precognition of Y', into two sets, viz. (1) *positive*, and (2) *negative*. I will take them in order.

1. *Positive conditions*

(i) The first positive condition is that X should be *either* (*a*) a *single* human action or human experience (as in most sporadic cases); or (*b*) a *sequence*, $X_1, X_2, \ldots X_n$, of human actions (as in quotitative experimental cases), all realizing one or another of a certain limited set of alternatives, but some realizing some of these, and others realizing others of them. On the first alternative, Y will itself be a *single* event or state of affairs, and it may be of any kind. On the second alternative, Y will itself be a sequence, $Y_1, Y_2, \ldots Y_n$, all realizing one or another, and some one and some another, of the same alternatives as are realized by the X's.

(ii) The second positive condition is that Y shall still have been *future* when X was present. This takes a different form according as we are concerned with (*a*) a *single* X and a *single* Y, or (*b*) a *sequence* of X's and a *sequence* of Y's. On the first alternative, all that is necessary is that X should have taken place at a certain moment t_1, and Y at a certain later moment t_2. On the second alternative, terms such as X_r and Y_r, which occupy corresponding positions in their respective sequences, are simultaneous with each other. What we are concerned to compare is each successive X-term, e.g. X_r, with the Y-term which comes an assigned number of

[8] *Proc. S.P.R.*, Vol. XLIX, 1949–52.
[9] *Journal S.P.R.*, Vol. 41, No. 711, March 1962.

THE NOTION OF 'PRECOGNITION'

places *after* Y_r in the Y-series, e.g. Y_{r+p}. And the question is: Do X_r and Y_{r+p} realize the *same* one of the various alternatives, all of which are open to each of them?

(iii) The third positive condition is that Y should correspond to X in one or another of the various ways which I will mention below. Here, again, we must distinguish between the case of (*a*) *one single Y* and *one single X*, and (*b*) a *sequence* of *Y*'s and a *sequence* of *X*'s. We may cover both these cases by saying that the later event or state of affairs must correspond to the earlier in such a way that the later can be described as *fulfilling* the earlier. We can then distinguish the two cases by speaking respectively of *singular* and of *statistical* fulfilment.

Singular fulfilment would cover, e.g. the following kinds of correspondence between Y and X: (*a*) X is a felt impulse or an inner admonition to act or to abstain from acting in a certain way; and Y is a later event or state of affairs such that any normal person in the agent's position, if he had foreseen it, would have wished to behave as the impulse directed. (*b*) X is a dream or a waking hallucination as of perceiving such and such an event or state of affairs; and Y is exactly or predominantly such an event or state of affairs, which later came to pass. (*c*) X is a mental image or a dream or a waking hallucination; and Y is a later event or state of affairs such that X symbolically prefigured it.

Statistical fulfilment consists in there being a very substantial excess in the *actual* proportion of X_r's which are hits on Y_{r+p}'s over the proportion most probable on the hypothesis that such hits are merely chance-coincidences.

2. *Negative conditions*

What we have to consider under the head of 'negative conditions' is any circumstance under which we should *not* be prepared to say that X is a precognition of Y, even though all the positive conditions were fulfilled. The *absence* of each such circumstance is a negative condition for X to be a precognition of Y.

(i) The first negative condition is that it should not be a mere *chance coincidence* that X was followed by an event or state of affairs Y, so correlated with it that Y counts as a fulfilment of X.

(ii) In order to state the second negative condition it will be convenient to proceed as follows. Suppose that Y certainly was so correlated with X as to count as a fulfilment of it; and suppose, further, that we were quite certain that the occurrence of such a Y, after such an X had happened, was not a mere chance-coincidence. Under what circumstances should we still be inclined to deny or to doubt that X was a precognition of Y?

If the admitted facts of a case of ostensible precognition could be certainly or plausibly explained in any one, or any combination, of the following ways, we should decline or hesitate to call it a case of *genuine* precognition. The alternative explanations about to be mentioned have been listed by careful writers on precognition, such as Mrs. Sidgwick, Saltmarsh, and Mr. Fraser Nicol. They may be divided into (*a*) those involving nothing but *normal* factors; (*b*) those involving factors which are *abnormal*, but not paranormal; and (*c*) those involving *paranormal* factors.

Under (*a*) come cases where the fulfilment might be due to the subject's own voluntary action, or to auto-suggestion, or to suggestions conveyed to him normally by others. Under (*b*) come cases where the fulfilment might be due to one or other or both of the following causes, viz. (α) the acquirement by the subject of relevant data through an abnormal acuity of his ordinary senses; and/or (β) the possession by him of abnormally developed powers of subconscious reasoning, of numerical calculation, or of non-inferential bodily adaptation, applied by him to the data at his disposal.

The contents of (*c*) are, naturally, a very mixed bag. They range from explanations in terms of well-attested paranormal powers, exercised by ordinary persons in the flesh, to explanations which postulate the existence and operations of surviving spirits of dead men, or of non-human intelligent beings (embodied or unembodied), and which ascribe to these almost miraculous powers of cognition and action.

An example of the least spectacular kind would be any case of ostensible precognition by A of an action subsequently performed by B, which might plausibly be accounted for by supposing that B had already formed a conscious or unconscious intention to perform that action, and that A became aware of that intention by simultaneous telepathy. Another conceivable explanation, in terms of telepathy, which might apply to a case where A ostensibly precognizes an action subsequently performed by B, would be this. We might suppose that A had unwittingly formed the intention that B should perform such an action, and that this unconscious desire of A's acted telepathically on B as a kind of hypnotic suggestion.

To go rather farther, suppose we were prepared to accept clairvoyance as a well-attested paranormal accomplishment. Then some cases of ostensible precognition might plausibly be explained by supposing that the subject had unconscious clairvoyant awareness of certain contemporary physical events and states of affairs, and that he unconsciously inferred from this and from his knowledge of the relevant laws of nature that certain physical developments would take place. Venturing still farther into the preposterous, we might ascribe to the subject telekinetic powers, whereby he could act, without using his hands or other limbs, on remote bodies, including, perhaps, the brains and nervous systems of other men. On that assumption many cases of ostensible precognition might be explained by ascribing the fulfilling event to the subject's unconscious telekinetic action.

So far I have confined myself to examples of paranormal explanations of ostensible precognition which do not postulate the existence of any intelligent agents other than ordinary human beings in the flesh. But suppose one holds that there is good evidence that certain human beings have survived bodily death and that they can communicate through mediums. Then, if a prediction as to the future action of someone still alive in the flesh were to be received through a medium in the form of an ostensible communication from the surviving spirit of a certain deceased human being, and if it were to be fulfilled, that might quite plausibly be explained on the following lines. One might suppose that the living person in question had already subconsciously formed an intention to act in the way predicted; that the surviving spirit was telepathically aware of that intention, and had inferred in the normal way that it would probably be carried out; that the prediction was simply a communication to the sitter, through the medium, of the conclusion of that inference; and that the fulfilment was due simply to the person concerned carrying out his intention in the normal way.

I have now considered a number of different lines on which one might try to explain a case of ostensible precognition. Just in so far as any such explanation of a case was held to be plausible, it would be held to be *doubtful* whether it could properly be counted as a case of *precognition*. The question now is this: Can all such explanations be brought under a single head? If so, we may take the denial of the possibility of any explanation falling under that head as the second negative condition which a case must fulfil if it is to count as a case of *precognition*.

Now I think that all such explanations as we have been considering do fall under one certain heading. I will consider first the case where X is a single event, and Y is a single later event or state of affairs which fulfils it. After that I will deal with the case where X and Y are sequences of events, and where the fulfilment is statistical.

When we are concerned with a *single X*, and a *single* later *Y* which fulfilled it, all the explanations can be brought under the following head: *Either* (*a*) X contributed (whether immediately, or through an intermediate chain of effects which were also causes) to cause a later event Y, of such a kind as to count as a fulfilment of X; *or* (*b*) though X did *not* contribute to cause Y, there was an event or state of affairs W, earlier than both X and Y, which contributed to cause first X and then later on such a Y as would count as a fulfilment of X. On the first alternative, we may say that there is *a single causal chain*, in which X is a cause-factor in an earlier link and Y is an effect-factor in a later link. On the second alternative, we may say that there are *two causal chains, diverging from a common link in which W is a cause-factor*; that X is an effect-factor in a later link in one of them (say C_1); and that Y is an effect-factor in a still later link in the other (say C_2).

Our second negative condition, where we are concerned with a *single X* and a *single* fulfilling Y, would therefore be the *denial* that any such explanation is possible in the case in question. It may be stated as follows: The occurrence, after X, of such an event or state of affairs as Y, *cannot* be explained, either (*a*) by X having been a cause-factor in a causal ancestor of Y, or (*b*) by X and Y being causal descendants, in different lines of causal ancestry, from a common ancestral cause-factor W.

Let us now consider the case where X is a *sequence*, $X_1, X_2, X_3, \ldots, X_n$, and Y is a *sequence*, $Y_1, Y_2, Y_3, \ldots, Y_n$; and where the fulfilment is *statistical*. Those who wish to exclude precognition have got to explain, on the same general principles as above, the fact that the proportion of X_r's which are hits on Y_{r+p}'s exceeds the proportion most probable on the hypothesis of chance-coincidence to a degree which is highly improbable on that hypothesis.

Explanations on the above lines would take the following alternative forms. (a) We might suppose that either *each* X_r, or an *appreciable proportion* of X_r's, influence (whether immediately or through a chain of intermediate causes and effects) the nature of Y_{r+p}, in such a way as to make it tend to conform to X_r. (If that influence be exerted by *each* X_r, it is plain that it must be weak enough to be overcome in the case of many of them. If, on the other hand, it be exerted by *only some* of them, we may suppose it to be as strong as we like, provided that the proportion of such effective X_r's be not too great.)

(b) Instead, we might suppose that the two members, either of *each* couple X_r and Y_{r+p} or of an *appreciable proportion* of such couples, are causal descendants, in different lines of causal ancestry, of a common causal ancestor W_r (different for each such couple); and that W_r tends to produce first an X_r of a certain kind and in the one causal line, and later a Y_{r+p} of *the same kind* in the other causal line.

So the second negative condition, where we are concerned with *sequences* and *statistical* fulfilment, is that *no* explanation on the lines of either (a) or (b) is possible in the case in question.

Subject to the above detailed explanations, we may state the second negative condition roughly and briefly as follows. The fulfilment cannot be explained *either* (a) by the later event being a causal descendant of the earlier one, *or* (b) by their being both causal descendants, in different lines of causal ancestry, of a common causal ancestor.

3. *The definition*

We are now, at long last, in a position to define the statement 'X was a precognition of Y'. The definition would consist of the following five clauses:

(i) X was either (a) *a single human action or human experience*; or (b) *a sequence of human actions*, $X_1, X_2, \ldots, X_n$, all realizing one or another of a certain limited set of alternatives, and some realizing some, and others realizing others, of these. In case (a) Y was a *single* event or state of affairs *of any kind*. In case (b) Y was a *sequence*, $Y_1, Y_2, \ldots, Y_n$ of events or states of affairs, all realizing one or another, and some one and some another, of the same alternatives as are realized by the X's.

(ii) In case (*a*) X happened at a certain earlier moment t_1, and Y at a certain later moment t_2. In case (*b*) terms, such as X_r and Y_r, which occupy corresponding positions in their respective sequences, are simultaneous with each other. What has to be considered, in relation to a typical term X_r in the X-sequence, is the term Y_{r+p}, which comes an assigned number of places ahead of Y_r in the Y-sequence. And the question is whether X_r and Y_{r+p} do or do not realize the same one of the various alternatives open to each of them. If and only if they do, we say that X_r was a *fore-hit* on Y_{r+p}.

(iii) In case (*a*) Y corresponded in detail to X, in one or another of the various ways appropriate to the various forms which X may have taken, so that we should say that Y was a *fulfilment* of X. In case (*b*) the Y_{r+p}'s in the aggregate corresponded to the X_r's, in that the *actual* proportion of $+p$ fore-hits *exceeded the proportion most probable on the hypothesis of chance-coincidence* to an extent which is highly improbable on that hypothesis. Here we talk of *statistical fulfilment*.

(iv) In case (*a*) it was *not a mere chance-coincidence* that X was followed by a Y, so correlated with it that Y counts as a fulfilment of X. In case (*b*) it was *not a mere chance-coincidence* that the proportion of X_r's which were hits on Y_{r+p}'s should have so greatly exceeded the proportion most probable on the hypothesis of chance-coincidence.

(v) Finally, the occurrence in case (*a*) of a Y, so related to X as to be a fulfilment of it, *cannot* be accounted for, either (α) by X being a cause-factor in a causal ancestor of Y, or (β) by X and Y being both causal descendants, in different lines of causal ancestry, of a common ancestral cause-factor W. And, in case (*b*), the occurrence of so improbably large an excess over chance-expectation in the proportion of X_r's which are fore-hits on Y_{r+p}'s *cannot* be explained by appropriate hypotheses on the lines of (α) or (β) above.

DIFFICULTIES IN THE WAY OF ADMITTING PRECOGNITION

Having defined the statement that X was a precognition of Y, we are now in a position to consider the difficulties which have been felt to lie in the way of admitting the existence of precognition. These may be divided into (1) *Empirical*, and (2) *A priori*. The latter seem to me to be much the most interesting and important. I will take the two in turn.

1. *Empirical*

Even if there be nothing impossible in the very notion of X being a precognition of Y, as defined above, it would evidently be extremely difficult to be sure that any alleged case, however well attested, really answered to the definition. The practical difficulties arise in connection with the two *negative* conditions, stated in clauses (iv) and (v) of the definition.

In the first place, there is no objectively valid criterion for deciding whether an antecedently improbable sequence of one event on another was so unlikely that it cannot reasonably be regarded as a mere chance-coincidence. Take, e.g. Mrs. C's fulfilled dream of being pursued by a monkey. It is an extremely rare event for a monkey to be loose at any time in any street in London. It is a fairly rare event, even for a person who has a morbid dislike of monkeys, to have so impressive a dream of being pursued by one that she mentions the dream next morning and is advised to take a walk to dispel the unpleasant impression. It is a most extraordinary coincidence that such a rare event as a monkey being loose in a London street should have happened, on just the morning after Mrs. C had dreamed of being pursued by a monkey, and in just the street in which she was taking a walk. Many people will find it impossible to regard this as a mere chance-coincidence; many will find no difficulty in doing so; and, once they are agreed as to the relevant statistical frequencies, there is no further room for rational argument between them on the point at issue.

Suppose, however, that one *is* persuaded, in regard to a certain case, that the sequence of such a Y on such an X was too improbable to be a mere chance-coincidence. Then one is faced with a second difficulty. It is impossible to be sure that *no* explanation, on the lines ruled out by clause (v) of our definition, could be given in terms of normal, abnormal, or the wildest kinds of assumed paranormal powers, on the part of persons still in the flesh, or of the surviving spirits of dead men, or of supposed non-human rational beings.

It may be objected that no sensible person need hesitate to exclude explanations on the lines ruled out by clause (v), if these have to resort to postulating paranormal powers for the like of which we have no independent evidence, or entities in the existence of which we have no independent grounds for believing. The fact that some quite sensible persons *do*, nevertheless, put forward such explanations is undoubtedly due to their holding, explicitly or unwittingly, that there is an insuperable *a priori* objection to the very notion of precognition. Suppose that one holds that view, and suppose that one is convinced that there are well-attested cases of ostensible precognition which are not just matters of chance-coincidence. Then there is nothing for it but to explain them, by hook or by crook, on the lines ruled out by clause (v); and to swallow any factual assumptions, however fantastic, that such an explanation may require. So we may pass now to the *a priori* difficulties.

2. *A priori*

I think that there are two theoretical difficulties which have been held to inhere in the very notion of precognition. It is likely that they have not always been very clearly distinguished. I will describe them as (i) the *Epistemological Difficulty*, and (ii) the *Causal Difficulty*. I believe that the first is illusory, and the second is very serious indeed. I will now consider them in order.

(i) *The alleged Epistomological Difficulty*. The alleged epistemological difficulty may be put as follows. A *precognitive* experience would, from its very nature, be, or would involve as an essential factor, a *present* state of *direct acquaintance* with a *still future* event or state of affairs. But at the time when the allegedly precognitive experience was occurring there *was not as yet* any such event or state of affairs to be its object; and at the time when that event or state of affairs came to be there was *no longer* that experience. So nothing could possibly fulfil the requirements of a genuine precognition.

I think that this objection rests on the following two assumptions, viz. (*a*) that precognition, if it were possible, would be epistemologically of the nature of *perception*, i.e. that it would be quite literally '*pre-perception*'; and (*b*) that an ordinary perception is a state of *direct acquaintance with* the thing or event or state of affairs

THE NOTION OF 'PRECOGNITION'

of which it is said to be a perception. Neither of these assumptions is tenable.

As regards (*a*), it is probably true that most experiences which have turned out to be *prima facie* precognitive have resembled perceptions, at least in the respect that they were *intuitive*, and not merely discursive, in character. A minority of them have been waking hallucinatory *quasi*-perceptions, in which the subject had an experience as of seeing such and such things and persons and events, as of hearing such and such sounds, etc. And a majority of them have been dreams, in which the dreamer's experience is essentially similar in kind to hallucinatory *quasi*-perception in the waking state. And many, which were neither waking hallucinatory *quasi*-perceptions nor dreams, at any rate involved as an essential factor the immediate awareness of a *mental image*, visual or auditory, recognized as such at the time.

Now a person may have a dream, or a waking hallucination, which can be shown to correspond very closely to a certain scene and certain events in it which he witnessed in the *past*. No one, in *that* case, would be inclined to say (except in an admittedly metaphorical sense) that the experient had a *post-perception* of the past scene and the past events. Again, if a person now has a mere visual or auditory image, recognized by him as such, and if this were found to correspond very closely to something which he had seen or heard in the *past*, we should be even less inclined to call his present image-experience a *post-perception*. If we did so, we should admittedly be talking metaphorically.

Now, to say that a waking hallucination or a dream or an imaginal experience was 'fulfilled' is to say neither more nor less than that a certain *later* event or state of affairs was found to correspond to it in the kind of way in which a certain *earlier* event or state of affairs is often found to have corresponded to such an experience. Since, in the latter case, we should not think of regarding the experience as *literally* a post-perception of the corresponding past occurrence, there can be no reason for regarding the experience, in the former case, as *literally* a pre-perception of the corresponding future occurrence. Of course, there is no harm in talking of it as a 'pre-perception', provided that one understands and makes clear to others that one is using that term *metaphorically*. But, in that case, one is not entitled to draw any of the inferences which might be justified if one were using the term 'pre-perception' literally.

I pass now to (b), viz. the assumption that, if an experience is *literally* a perception, then it must be a state of direct acquaintance with the thing or event or state of affairs of which it is said to be a perception.

That this is a mistake can be shown most readily by considering cases where an experience, which is quite literally and correctly described as a perception of a certain event, does not begin until well after that event has ceased. Suppose, e.g. that a gun is fired on a certain occasion, when the air is still and I am at a distance of one mile from it. In due course I shall have an auditory experience of a characteristic kind, which would correctly be described as 'hearing the explosion'. But that experience would not have begun until 4·84 seconds after the event heard had ceased. Suppose, again, that an observer is watching the Sun from the Earth, and that at a certain moment he begins to have an experience which would be correctly described as 'seeing a bright eruption of flame in the neighbourhood of a sunspot'. The event thus seen will have begun some eight minutes before the visual perception of it begins. Suppose, then, that it should last only for four minutes. Then the process perceived as a bright eruption of flame on the sun will have *ceased* four minutes before the experience, which is the observer's visual perception of it, *begins*.

The utmost that can be admitted, then, is that, when one has a perceptual experience, one *uncritically takes for granted* that what one is perceiving is *simultaneous with it*. Since it is certain that in many cases what one is perceiving has ceased to exist *before* the perception of it begins, it is certain that this uncritical taking for granted of simultaneity is often mistaken. Therefore, it cannot be made the basis of a denial in principle that a perceptual experience might *quite literally* be a perception of something which will not begin to exist until after the experience shall have ceased.

The alleged epistemological difficulty, therefore, vanishes in smoke. In the first place, it is very doubtful whether any pre-cognitive experience is *literally* a pre-perception of the event or state of affairs which will in due course fulfil it. And precognition, in so far as it is not literally of the nature of perception, is *epistemologically* on all fours with ordinary non-inferential retro-cognition, which admittedly presents no particular epistemological difficulty Secondly, even if some precognitive experiences were literally pre-perceptions, no *epistemological* objection in principle could legitimately be based merely on the ground that the object of a perceptual experience *must* exist simultaneously with that experience. For we know, from such examples as I have given, that that is not true.

(ii) *The causal difficulty*. We pass, finally, to what I regard as the really serious *a priori* difficulty, viz. the causal one.

Let us begin by comparing and contrasting (*a*) an ostensible pre-perception (e.g. a waking hallucination, in which the subject ostensibly sees a certain friend knocked down by a certain kind of car in front of a certain shop in a certain street in London, and where, a few hours later, that friend is in fact knocked down by exactly that kind of car in front of that very shop in that same street) with (*b*) an ordinary auditory post-perception of the discharge of a distant gun, or an ordinary visual post-perception of a bright eruption of flame in the neighbourhood of a sun-spot. We at once notice the following profound difference.

In the cases of post-perception, the past event is connected with the present perceptual experience by a *causal chain* of successive events, each an effect of its immediate predecessor and a cause of its immediate successor, initiated by the event perceived and leading up to the event which is so-and-so's post-perception of that event. In a case of pre-perception, if such were possible, there *could* be no analogy to this. *Until* an event, which will answer to the present experience in such a way as to make that experience count as a pre-perception of it, shall happen, *nothing can be caused by it*. Therefore that event *cannot* have contributed, either directly or through a causal chain of intermediate events, to cause the experience which is said to have been a pre-perception of it.

Let us next compare and contrast (*a*) a dream or a waking bit of imagery, which is found to correspond with a certain *past* experience had by the subject or with a certain scene which he has witnessed or taken part in *formerly*, with (*b*) a dream or waking bit of imagery, which turns out to have corresponded with some *later* event or state of affairs, in such a way that we say that the latter 'fulfils' the former. Here we should not call the experience in the one case *literally* a post-perception, and we should not call it *literally* a pre-perception in the other. But the results of the comparison are essentially the same as those which we reached above.

In the case of a *retro*-referential dream or bit of waking imagery, we secure causal continuity by making an assumption of the following kind. We assume that, when a person has an experience, it sets up a 'trace' in him; that this persists indefinitely (e.g. as a structural modification, or as a continually repeated cyclical process of some kind or other) in some part of his brain or nervous system; that such a trace can from time to time be 'excited' by new experiences, or by internally initiated changes, bodily or mental; and that, when excited, it tends to evoke an experience somewhat like that which initially set it up. No doubt, this kind of story contains a good deal of myth. But it does at least tell us of something which fills the temporal gap between the dead and gone past event and the present experience which refers back to it, viz. something persistent, which the past event contributed immediately to cause, and which now contributes immediately to cause the present experience. The fact that we are willing to swallow so much mythology here shows how strong is the felt need for *some* such causal filling of the temporal gap.

Now, plainly, nothing in the least analogous to this is possible in the case of a *pro*-referential dream or bit of imagery and its fulfilment. We have explicitly ruled out, by definition, the suggestion that it contributes to set up a chain of causes and effects which will eventually produce the event or state of affairs which will fulfil it. And the alternative suggestion, viz. that the *fulfilling event or state of affairs* contributes to set up a chain of effects and causes which contributes to cause the *pro-referential experience*, is plainly nonsensical. For, *until* the event which will answer to the present experience in such a way as to be a fulfilment of it, shall have happened, *nothing* can be caused by it. And, *when* it shall have happened, anything that it may contribute to cause must be *later than it*.

I would like to make it quite clear at this point that the difficulty does not arise from the *purely linguistic* fact that the words 'cause' and 'effect' in English connote that a cause precedes its effect; so that it would be a *contradiction in terms* to talk of an effect preceding its cause. If that were all, we could easily deal with the difficulty by one or other of the two devices which natural scientists and mathematicians have repeatedly used in such circumstances, viz. either by continuing to use the old word, but explicitly giving to it an extended technical meaning to cover the new facts, or by introducing and defining a special new technical term to replace the old word in the new context.

THE NOTION OF 'PRECOGNITION'

The real source of the difficulty is not linguistic, but is *factual*. It is the self-evident fact that what we call a 'future event or state of affairs' *is* nothing but an unrealized possibility, until it happens or 'comes to pass'; and that which *is not* cannot possibly *do* anything, and therefore cannot be a factor *influencing* anything. Very likely it is this self-evident non-linguistic fact which has moulded the linguistic convention governing the use of words like 'cause' and 'effect', and has thus rendered *self-contradictory* any sentence in which an attempt is made to assert the occurrence of what I may call 'retrogradient causation', i.e. causation of what is now happening by something which has not yet happened.

I think that the essential difficulty tends to be slurred over, if we look backwards from the present moment t_3 to two events e_1 at t_1 and e_2 at t_2, both in the past. There is then, perhaps, some temptation to say: 'After all, why should not the later event have contributed to cause the earlier, just as an earlier event often contributes to cause a later?' What we have to bear in mind, in face of such temptation, is this. To say that a *later* event influenced an *earlier* one implies that an event, which at t_1 was *still future*, contributed towards determining the occurrence of an event of such and such a kind at t_1. Now the phrase '*future* event' does not describe an event of some special kind, as the phrase 'sudden event' or 'unfortunate event' or 'historic event' does. Suppose, e.g. that I refer now to my own death as a 'future event'. I am merely saying that *there will some day* be an occurrence correctly describable as 'the death of C. D. Broad'. Until that day shall arrive 'my future death' *is* nothing, and therefore can influence nothing; though, of course, the present *knowledge*, by myself and by others, that there will be such an event in the not very distant future, can and does influence my actions and theirs, on occasions when it is called to mind and is relevant.

It seems to me self-evident, then, that the later event or state of affairs, which is found to 'fulfil' an earlier experience and to make it *prima facie* precognitive, cannot possibly have contributed in any way, directly or indirectly, to determine the occurrence of that experience.

How, precisely, does that create a difficulty in the very notion of precognition? It does not do so at the first move. For (as the reader will see, if he refers back to the definition on p. 185) it is no part of the *definition* of 'precognition' that the fulfilling event or state of affairs should contribute to determine the occurrence of the experience which is said to be a precognition of it. The difficulty arises at the second move, and it does so in connection with the two negative clauses, (iv) and (v), of the definition.

According to clause (v), the occurrence of the later event Y, so correlated with the earlier experience X as to count as a fulfilment of it, *must not* be accountable for, either (*a*) by X being a cause-factor in a causal ancestor of Y, or (*b*) by X and Y being both causal descendants, in different lines of causal ancestry, of a common ancestral cause-factor W. To this we now have to add the self-evident fact (*c*) that Y cannot have contributed in any way to determine the occurrence of X. So clause (v) of the definition, and the self-evident fact just stated, together entail that, if X is to count as a precognition of Y, there can be no influence, direct or indirect, either of X on the occurrence of Y or of Y on the occurrence of X, and that X and Y cannot both be causal descendants, in different lines of causal ancestry, of a common ancestral cause-factor W.

But this seems *prima facie* to be incompatible with clause (iv) of the definition. For that states that, if X is to count as a precognition of Y, it *must not be a mere chance-coincidence* that X was followed in course of time by a Y, so correlated with it as to count as a fulfilment of it. Now, if there be *no* influence of X on the occurrence of Y or of Y on the occurrence of X, and if X and Y be *not* both causal descendants, in different lines of causal ancestry, of a common ancestral cause-factor, how can the sequence of such an event as Y on the event X be *anything but* a mere chance-coincidence?

CONCLUSION

So far as I can see, then, there could not possibly be a case of genuine precognition, as defined by me. Anyone who thinks that there could be must do one or another of the following three things. Either (*a*) show that my definition is defective in regard to one or the other or both the two negative clauses (iv) and (v). Or (*b*) show that there is a sense in which it could properly be said that the sequence upon X of a Y, so related to X as to count as a fulfilment of it, is *not a mere chance-coincidence*, even though X has no causal influence on the occurrence of Y and Y has none on the occurrence of X, and X and Y are not both causal descendants, in different lines of causal ancestry from a common ancestor W. Or (c) that it *is* intelligible to talk of the occurrence of an event X at t_1 being causally influenced by an event Y, which had not then happened (and therefore was a mere future possibility), and did not happen (and so become actual) until t_2. (May I add that it would not be enough to cite eminent physicists who talk as if they believe this? What is nonsense, if interpreted literally, is no less nonsense, if so interpreted, when talked by eminent physicists in their professional capacity. But when a way of talking, which is nonsensical if interpreted literally, is found to be useful by distinguished scientists in their own sphere, it is reasonable for the layman to assume that it is convenient short-hand for something which is intelligible but would be very complicated to state in accurate literal terms.)

Suppose, however, for the sake of argument, that I should be right in thinking (a) that my definition of 'X was a precognition of Y' makes explicit what most people have at the back of their minds when they use, and when they hesitate or decline to use, that phrase; and (b) that nothing could possibly answer to *all* the clauses of that definition. Then, when we are talking carefully, we shall have to confine ourselves to the phrase '*ostensible* precognition', and to bear in mind that no case of ostensible precognition can possibly be one of genuine precognition.

That, however, will do no harm to us who are concerned with psychical research. The important thing to get out of our minds is that, if there were precognition, it *must* be of the nature of *pre-perception* of the future fulfilling event or state of affairs; and that if it be of the nature of perception, it must, for that very reason, be a state of present *direct acquaintance with* the as yet future fulfilling event or state of affairs. Both parts of this assumption are certainly false, and they are the source of many pseudo-problems which bedevil the subject.

Having rid ourselves of that superstition, we shall be left with clearer minds for dealing with any well-attested cases that may fairly be counted as cases of *ostensible* precognition. These will be cases where (a) the kind and degree of correlation between the earlier X and the later Y is so great that we hesitate to regard it as mere chance-coincidence, and (b) there is no plausible causal explanation of it in terms of generally admitted normal human capacities, even when present to an abnormal degree of sensitivity and efficiency. All such cases will remain of great interest and importance to the psychical researcher. It will be his business, with the aid of any other scientists who are willing to help him, to suggest causal explanations, in the first instance in terms of *paranormal* powers which are already admitted or strongly suspected to exist in some human beings alive in the flesh, and to have manifested themselves in phenomena *other than* ostensible precognition.

VIII

THE EXPLANATION OF ESP

C. W. K. Mundle*

Nearly all philosophers and scientists who have tried to explain ESP have offered us theories applicable to telepathy, but not applicable to—and indeed incompatible with the occurrence of—clairvoyance; and those who have offered such theories have acknowledged clairvoyance with little more than a footnote or a parenthesis. Yet the evidence for clairvoyance is surely as strong as that for telepathy.

Most of Professor J. B. Rhine's star subjects in the early 1930s scored equally well in telepathy and clairvoyance conditions, and the successful clairvoyance experiments at Duke University included some of the best controlled series, e.g. the Pearce–Pratt experiment,[1] and many of Tyrrell's experiments with Miss Johnson were done in clairvoyance conditions.[2] If we allow the possibility of precognition, it is necessary to eliminate the possibility of precognitive telepathy in a 'pure clairvoyance' experiment, and the possibility of precognitive clairvoyance in a 'pure telepathy' experiment.

There is, I think, only one 'pure clairvoyance' experiment which has yielded decisively high scoring, i.e. Tyrrell's series in which the commutator and the delay-switch were both in use; but there

---

* Reprinted from the *International Journal of Parapsychology*, Vol. VII, No. 3, Summer 1965.

[1] See 'Some Basic Experiments in Extra-Sensory Perception' (author not identified), *J. Parapsychol.*, Vol. I, No. I, March 1937; and J. B. Rhine and J. G. Pratt, 'A Review of the Pearce–Pratt Distance Series of ESP Tests', *J. Parapsychol.*, Vol. 18, No. 3, September 1954.

[2] G. N. M. Tyrrell, 'Further Research in Extra-Sensory Perception', *Proc. S.P.R.*, Vol. XLIV, 1936–7.

is only one 'pure telepathy' experiment which has yielded decisively high scoring, i.e. Soal and Bateman's with Mrs. Stewart.[3] I am deliberately ignoring experiments which have yielded odds against chance of the order of one or two hundred to one. This may appear high-handed, but it seems to me that the Spencer Brown controversy[4] should have taught us that probability values of this magnitude are not sufficient in a crucial experiment, e.g. one designed to isolate a certain species of ESP, particularly when we do not know how many unsuccessful experiments of the same type may have been carried out but not published. I shall describe briefly each of the crucial experiments to which I have referred.

Tyrrell's pure clairvoyance experiments

The subject, Miss G. M. Johnson, sat beside five small light-tight wooden boxes, each containing a tiny electric lamp. Her task was to lift the lid of the box in which the lamp had just been (or was about to be) lit. The operator, Tyrrell, screened from the subject, switched on a lamp by pressing one of five keys. The electrical connections between the keys and the lamps were 'scrambled' by a commutator which was enclosed in a box, so that when this box was closed no one could tell by observation what these connections were. The result of each trial was recorded automatically by a line drawn on a moving tape. A line was drawn when any *one* (but not if more than one) of the boxes was opened. When the correct box was opened, a second line was drawn on the tape parallel to the first. *The record on the tape gave no indication which box had been opened at any trial.*

In view of this, precognitive telepathy could not have helped the subject. Suppose that she was able to discover by precognitive telepathy—from the visual experience of someone who later in-

[3] S. G. Soal and F. Bateman, *Modern Experiments in Telepathy* (London: Faber, 1954), pp. 255–8.

[4] See C. D. Broad, *Lectures in Psychical Research* (London: Routledge, 1962), pp. 74–91; D. Spencer Brown, Letters to *Nature*, 25 July 1953 and 26 September 1953; *idem*, *Probability and Scientific Inference* (London: Longmans, 1957), especially Appendices I and II; C. W. K. Mundle, 'Probability and Scientific Inference', *Philosophy*, Vol. XXXIV, 1959; J. Fraser Nicol, 'Randomness: The Background and Some New Investigations', *Journal S.P.R.*, Vol. 38, No. 684, 1955; and A. T. Oram, 'An Experiment with Random Numbers', *Journal S.P.R.*, Vol. 38, No. 683, 1955.

spected the tape—that her next trial was going to produce a hit, this could not help her to decide which box to open next. To eliminate the possibility that Miss Johnson might be succeeding by hyperaesthesia to heat (or other) radiation from the lighted lamp, Tyrrell introduced a delay-switch. The operator's pressing a key then merely determined that a certain lamp *would* light *if and when* the correct box was opened. Under these conditions (with the commutator concealed and the delay-switch in use), Miss Johnson, in a series of 855 trials, scored at her usual level—26 per cent against a mean chance expectation of 20 per cent ($p =$ nearly 10^{-6}). These experiments were carried out in 1936.

Soal and Bateman's pure telepathy experiment

The subject was Mrs. Gloria Stewart and the agent was Soal himself. The agent thought of one of the five target-symbols, taking them in an order which was jointly determined by (*a*) a random series of the digits 1 to 5 which was objectively recorded, and (*b*) a *private* code linking each digit with one of the symbols (i.e. *private* in the sense that Soal never recorded it or formulated it in spoken words). The subject's guesses were scored by the agent, who recorded only the total number of hits on each run. He did not mark the correct guesses, or let his eyes linger on them, since this would, in theory, have permitted the subject to identify the code by precognitive clairvoyance! So that we should not have to rely only upon his own memory and testimony, Soal got his collaborator, Bateman, who was an old friend, to *infer* the code he was using by making indirect allusions to their past shared experiences, e.g. jokes about mutual friends, to check and confirm Soal's scoring. In the 40 runs done with this method, Mrs. Stewart averaged 7·3 hits per run, a little better than her usual level of success ($p = 10^{-11}$). This experiment was carried out in 1946.

Before going any further I must define my use of the terms 'telepathy' and 'clairvoyance'. I reject the familiar but question-begging definition of 'telepathy' as the transfer of information from one person's mind to another's. If we define 'telepathy' thus, it is impossible to verify whether there has ever been a case of telepathy; for when, in so-called telepathy experiments, information possessed only by the agent is acquired by the subject, we have no way of verifying whether this information originated from the *mind* of the agent or from his *brain*. To make the occurrence of telepathy verifiable, we *must* define 'telepathy' as: ESP in which the source of the information is *another person* (whether from his mind *or* his brain—we must leave this open). We must then define 'clairvoyance' as ESP in which the source of the subject's information is physical objects other than the brain (or central nervous system) of another person.

It has been argued by many distinguished people, including Professor H. H. Price,[5] that ESP cannot be explained by Physics. Therefore, they continue, this disproves Materialism and thereby establishes, or at least supports, the case for Psychophysical Dualism (though not necessarily Descartes' version, which Price rejected). Those who argue thus, including Price, usually start with the question-begging definition of 'telepathy'. Their main case is that ESP in order to be explained by Physics, on the analogy of radio transmission, must be attributed to a physical radiation, and that this theory breaks down. The stock arguments for this conclusion are as follows:

1. The Inverse Square law applies to all known forms of physical radiation (unless the radiation is focused in a beam, when there may be very little attenuation with distance; but this does not seem relevant since there have been successful ESP experiments with a group of subjects located in different directions). On the physical radiation theory, it is argued, we should find a marked decline in scoring level if the subjects' distance from the source of radiation is considerably increased. But all investigators agree that there is no covariation between distance and scoring level, and successful experiments have been done over distances of several hundred miles. This argument, by itself, is inconclusive. The critic may reply: ESP is being assimilated to radio transmission; radio sets have amplifiers and volume controls; we may suppose then that the brain possesses an automatic volume control which amplifies weak ESP signals. (The many misses in ESP tests must then be ascribed to factors other than weakness of the signals, but this can be done.)

2. Physical barriers do not affect ESP, as one would expect on the radiation theory. The most thorough experiments yet done on screening were at the Institute of Physiology at Leningrad by

[5] See H. H. Price, 'Psychical Research and Human Personality', *Hibbert Journal*, Vol. XLVII, 1948-9, and Ch. 3 of this volume.

THE EXPLANATION OF ESP

L. L. Vasiliev.[6] The Russian research team set out with the assumption that ESP must be due to electro-magnetic radiation. To verify this, their subjects or/and agents were completely enclosed in metal containers which completely blocked all electro-magnetic waves between 1 millimetre and 1 kilometre. ESP was not thereby impaired. Vasiliev concedes that there are very strong grounds for rejecting the hypothesis that ESP is due to electro-magnetic waves outside this range.

3. Electrical 'brain-waves' are too weak, according to W. Grey Walter, an authority on brain physiology.[7] Discussing the radiation theory of telepathy, he writes: 'If we consider the largest [electrical] rhythms of the brain as . . . radio-signals, we can calculate that they would fall below the noise level within a few millimetres from the surface of the head.'

These three objections taken together seem to eliminate the possibility of ascribing ESP to electro-magnetic radiation. Attempts have, however, been made to reconcile ESP with Physics by other means.

Unorthodox physicalist theories

1. The theory of Hans Berger, the physiologist who discovered the technique of recording brain-waves with the electro-encephalograph.[8] Berger recognized, like Grey Walter, that electrical changes in the brain are too weak to explain telepathy. He suggested that electrical energy in the agent's brain is transformed into 'psychic energy', which can be diffused to any distance, passing through obstacles without attenuation; that on reaching the subject's brain it is there transformed back into electrical energy, thereby producing neural patterns and experiences corresponding to those of the agent. Though Berger talked of 'psychic energy', he was not a dualist. He conceived of psychic energy as a form of physical energy, interchangeable with other forms.

2. The Theory of Ninian Marshall.[9] Like Berger, Marshall is

[6] L. L. Vasiliev, *Experiments in Mental Suggestion* (Church Crookham, England: I.S.M.I. Publications 1963). Trans. from the Russian.
[7] W. Grey Walter, *The Living Brain* (London: Duckworth, 1953).
[8] H. Berger, *Psyche* (Jena: Gustav Fischer, 1940). Reviewed by W. G. Roll in *J. Parapsychol.*, Vol. 24, No. 2, June 1960.
[9] N. Marshall, 'ESP and Memory: A Physical Theory', *Brit. J. Phil. Sci.*, Vol. X, 1959–60.

an epiphenomenalist, i.e. he assumes that all conscious experiences are determined by patterns of cortical activity. His theory of ESP is more revolutionary than Berger's, involving 'action at a distance'. He suggests that in telepathy the state of the agent's brain affects that of the subject, without there being any continuous chain of intervening events or indeed any transmission of energy. To render this intelligible, he offers his 'Hypothesis of Resonance', which states 'that any two substances exert an influence on each other which tends to make them become more alike. The strength of this influence increases with the product of their complexities, and decreases with the difference between their patterns'. This hypothetical force he labels 'the Eidopoic Influence'. He argues that this influence has hitherto produced observable effects only in the form of ESP, because the human brain is by far the most complex structure in the world, but he predicts that telepathic interaction will be detectable between computing machines if or when these can be made sufficiently complex in structure. This is a bold and intriguing speculation, though Price would perhaps be entitled to protest against Marshall's describing as a *physical* force one which involves no energy transfer. It appears that for physicists, physical transactions involve, by definition, a transfer of energy. Could Physics accommodate Marshall's revolutionary theory? (It is perhaps relevant that Professor Hoyle, the Cambridge physicist, has recently put forward a [not yet published] theory of gravity according to which gravity involves action at a distance.) We need not discuss this issue however, for it is time to recall the point stressed earlier: that the evidence for clairvoyance is as strong as that for telepathy. Berger and Marshall both ignore clairvoyance, and their theories must be rejected because they *preclude* clairvoyance. Both assume that the source of information in ESP is a brain, but in cases of clairvoyance there *is* no agent, no brain in which the relevant information has been recorded.

Is it possible for Physics to explain Clairvoyance?

There seems to be no hope of amending Marshall's theory to cover clairvoyance. He attributes telepathy to resonance between different brains on account of their structural complexity, but Zener cards are extremely simple in structure, so there should be no resonance between them and brains. Could Berger's theory be adapted? To do so, we would have to postulate the following: (*a*) that some form of physical energy is emitted not only by brains but by *all* physical objects (for it is not only cards which have been used in successful clairvoyance tests; we recall Tyrrell's experiments); (*b*) that the physicists (and they will find this hard to believe) have overlooked this energy in their theories and calculations. But when we try to advance further we meet formidable difficulties.

1. How are we to explain the subject's selectivity, his ability to discriminate the signals coming from the several target-objects? This problem is acute in the case of Rhine's 'Down Through' test, for the cards lie bunched together and one would expect signals emitted by the several cards to interfere with each other.

2. Whatever the nature of the hypothetical radiation, it must, since it penetrates screens, be very different from light-waves. But the information conveyed by clairvoyance concerns (*inter alia*) the *visible* properties of things, e.g. the shape or colour of a diagram. So we should have to assume that the radiation, though physically very different from light, happens to carry modifications corresponding to just those features of things which differentially affect light-waves and make us see different colours. A remarkable and unintelligible coincidence!

3. Physical barriers do not impede ESP; thus, unless we reject the principle of Conservation of Energy, we must suppose that only brains (or, at any rate, organisms) absorb and are affected by the hypothetical radiation, other physical things being, in effect, transparent to it. This would be an extreme paradox, since we have had to suppose that all physical things emit this radiation.

Enough has been said to show that *if* Price's thesis (that ESP disproves Materialism) were based on clairvoyance instead of telepathy, it would be too facile to dismiss it as merely an argument from our present ignorance. To explain clairvoyance, Physics would have to undergo a major revolution. Price would have been entitled to argue thus: 'the case for Materialism rests largely on scientific discoveries, notably the impairment of mental powers by brain damage, which make it plausible to claim that all mental powers can be explained in principle by Physics, by reference to transmission-processes and brain-mechanisms; therefore the case for Materialism is weakened by the fact that clairvoyance cannot be explained thus even in principle.' But Price claimed more: that

ESP disproves Materialism and supports Dualism. This goes too far. An Advocate for Materialism may reply as follows:

1. *Defence.* 'ESP cannot now be explained by Physics' does not entail 'ESP disproves Materialism', unless we define 'Materialism' as claiming *inter alia* that the current theories of Physics are correct and complete (and who believes this?). A materialist may define his position thus: that nothing exists except physical things and their states ('physical things' meaning those things which are both visible and tangible), and that conscious experiences are (or are inner aspects of) certain physical processes in higher organisms. 'Materialism' need not be so defined as to entail that physical things can interact only by contact or by means of physical radiations or fields. A materialist may regard the unobservables postulated by Physics (particles, waves, forces, etc.) as fictions, logical constructions, whose function is heuristic, to guide the scientists' imagination in formulating laws; so that what these laws are *about* is the observable behaviour of physical objects in the everyday sense of 'physical object'.

2. *Counter-attack.* How, we may ask, does Dualism help to explain ESP? Before we reject Materialism in favour of Dualism, because Physics cannot explain ESP, we are obliged to show that ESP *can* be explained in terms of non-material minds or mental forces; but have we any grounds for inferring that the unknown entities or forces responsible for ESP are of this nature?

This brings us to *Psychical Theories of ESP*. Dualists have offered various theories which are of about the same level of generality as the physicalist theories of Berger and Marshall. The commonest move is to postulate a collective unconscious mind, a concept expressed poetically by William James: 'there is a continuum of cosmic consciousness, against which our individuality builds but accidental fences, and into which our several minds plunge as into a mother-sea.[10]. The point of this move may be *either* (*a*) to suggest that one person's experiences may *directly* influence those of another, because all experiences of all people are really contents of one all-embracing Mind; *or* (*b*) to suggest that one person's experience can affect another's *indirectly*, via modifications in a substratum called 'the common unconscious.' (But

[10] Gardner Murphy and R. P. Ballon (eds.), *William James on Psychical Research* (London: Chatto, 1961), p. 324.

then we must ask what grounds there are for assuming that this substratum is *mental*.) We shall not survey the variation on this theme of 'the common unconscious', e.g. by Price (see Chapter 3), Tyrrell,[11] and Carington[12]; for such theories, when combined as they usually are with Dualism, are open to the fatal objection—that they can apply only to telepathy, not to clairvoyance. The only dualist theories which would cover clairvoyance seem to be the following:

1. Bergson's theory about normal memory and perception: that each person is potentially aware of all his past experiences and of all concurrent events, and that to prevent our being overwhelmed by useless knowledge, the brain suppresses all information except what is relevant to our present practical needs. It is a short step to equate the knowledge supposedly suppressed by the brain with ESP, and Bergson took this step.[13] It seems extravagant, however, to explain the sporadic exercise of ESP by a few people, by supposing that we are all potentially omniscient!

2. The theory of Thouless and Wiesner:[14] that when a person exercises clairvoyance (or psychokinesis), his mind is being influenced by (or is influencing) physical things outside his body *in the same way* as his mind normally interacts with his own brain. Cartesian Dualism is here simply taken for granted. Since this theory involves notorious problems concerning mind-brain interaction, the proposed account of clairvoyance and PK does not seem to me to render them more intelligible.

Many have argued that ESP provides evidence for Dualism. No philosopher seems to have argued that it counts in favour of Bishop Berkeley's form of Idealism, though this seems the stronger case, for this theory can accommodate the phenomena of clairvoyance. Berkeley's theory already involves telepathy, though not by this name. Berkeley held that material objects do not exist, if they are conceived as entities which exist independently of being perceived by minds. According to Berkeley, the objects (he called them 'ideas') which we see, touch, hear, etc., are produced in our

[11] G. N. M. Tyrrell, *Apparitions* (London: Duckworth, 1953), and *Science and Psychical Phenomena* (London: Methuen, 1938).
[12] Whately Carington, *Telepathy* (London: Methuen, 1945), see especially pp. 118–20.
[13] H. Bergson, Presidential Address, *Proc. S.P.R.*, Vol. XXVI, 1912–13.
[14] R. H. Thouless and B. P. Wiesner, 'The Psi Process in Normal and "Paranormal" Psychology', *Proc. S.P.R.*, Vol. XLVIII, 1946–9.

minds directly by volitions of the divine mind, i.e. by telepathy according to the traditional meaning of this term, i.e. the direct influence of one mind upon another mind. In order to accommodate our conviction that material things do continue to exist during periods when they are not being perceived by human minds, Berkeley argued that material things then exist as 'ideas' or 'arch-types' in God's mind. Thus the *phenomena* of clairvoyance could be accommodated by Berkeley's theory, and would then consist of more telepathy in which God played the role of Agent. In a successful clairvoyance experiment, the subject would be getting information about and from certain ideas in God's mind, for, on Berkeley's account, the concealed target-objects *consist* of God's ideas. I am neither an Idealist nor a Theosophist, but I feel obliged to offer them this argument for what it may be worth. But before they use it they should try to explain why the Deity should vouchsafe trivial information about card-symbols to a select few.

To sum up, if all ESP were reducible to telepathy (including precognitive telepathy), we already have several alternative theories (physicalist and dualist and idealist), which would render the facts more or less intelligible. If clairvoyance is an irreducible phenomenon, as it seems to be, the theoretical problems are *very* much more formidable. It seems to me possible to reformulate Physics to accommodate telepathy (as I have defined it), but well-nigh impossible to do this for clairvoyance. Parapsychologists have, perhaps, made a mistake in neglecting the problem of isolating pure telepathy and pure clairvoyance during the last eighteen years. One successful experiment in each is not enough. Before we can profitably try to explain ESP we need to know what are the hard facts, the irreducible species of ESP.

SUMMARY

It is argued that nearly all available theories about ESP apply to telepathy but not to clairvoyance; that the experimental evidence for clairvoyance is as strong as that for telepathy; that if we allow the possibility of precognitive telepathy and clairvoyance, we have for pure telepathy, and likewise for pure clairvoyance, only one really convincing experiment; and that to make the occurrence of telepathy verifiable, we must define 'telepathy' as: ESP whose source is the mind *or brain* of another person, and must then define 'clairvoyance' as: ESP whose source is physical objects other than another person's brain.

The familiar thesis—that ESP cannot be explained by Physics, thereby disproves Materialism and supports Dualism—is examined. The theory that ESP is transmitted by electro-magnetic radiation is open to objections which together appear decisive. Two unorthodox Physicalist theories are considered, but these apply only to telepathy and it seems impossible to adapt them to cover clairvoyance. Reasons are given for concluding that it is *extremely* difficult to reconcile clairvoyance with current Physics. Even so, the claim that ESP disproves Materialism goes too far, for (a) 'Materialism' need not be so defined as to imply that physical things interact only by contact or via physical radiations or fields, and (b) it seems no easier to explain ESP in terms of immaterial entities or forces. Dualist theories of ESP are briefly surveyed. Some of these could apply only to telepathy. Two such theories would accommodate clairvoyance, but they involve unresolved problems. The metaphysical theory which most easily accommodates both telepathy and clairvoyance is Berkeley's form of Idealism.

If all ESP is reducible to telepathy, including precognitive telepathy, we already have alternative theories, dualist and physicalist, which render the facts more or less intelligible. If clairvoyance is an irreducible species of ESP, the theoretical problems are much more acute. We need more evidence than we have of the occurrence of 'pure clairvoyance'.

IX

ESP IN THE FRAMEWORK OF MODERN SCIENCE*

Henry Margenau

As I look upon many researches in paranormal psychology and ESP, I am reminded of the state of affairs in physics some seventy years ago when scientists had first learned about the phenomenon of radioactivity. People might have gone around with Geiger counters or similar devices. They would have found them clicking here and there and everywhere—but just a little bit. On this evidence physicists surely could have tried to convince their colleagues that this was a real effect, that there was indeed something important in the idea of radioactivity—the emission of charged particles which was not understood by anyone at the time.

However, in those days physicists did not content themselves collecting samples and measuring them promiscuously. They did not go from one place to another making statistical studies of the intensity of this effect. Instead, they succeeded in finding circumstances under which the phenomenon became enhanced and became controllable. In other words, they were very specific, and by being specific, persistent, and judicious in their selection of experimental conditions they transcended the need for statistical arguments and probability estimates of validity; they uncovered clear evidence of the existence of the effect. My point is simply this: in order to study these obscure things, one must practise selectivity and concentrate one's attention upon instances where positive results are incontrovertible and, of course, demonstrably

* Luncheon address given at the ESP Forum sponsored by the A.S.P.R., 20 November 1965.

free of fraud. I suppose I am really doing nothing more here than echoing Dr. Pratt when he made a plea for the intensive study of high-scoring ESP subjects. I believe that as long as you go around making statistical studies everywhere and on everybody, you are not likely to be convincing for a long time to come.

Now the second thing needed, I believe, is theory. No amount of empirical evidence, no mere collection of facts, will convince all scientists of the veracity and the significance of parapsychologists' reports. He must provide some sort of model, to use Dr. Murphy's word; he must advance bold constructs—constructs connected within a texture of rationality—in terms of which ESP can be theoretically understood. The remainder of my paper will centre on possibilities of constructing such vehicles of understanding.

In order to put these problems in their proper context, let me say a few words about a matter which may at first seem irrelevant here, a few words about the competence and the limitations of science in general. Science is more than a mere collection of facts, a catalogue of observations. Observations alone, facts alone, lack the cohesion, the logical consistency, which every science demands. Science is a style of inquiry; it is a peculiar way of organizing human experience which integrates and thereby confers lucidity, clarity, and cohesion upon our immediate sense impressions and upon our observations. Every kind of human experience, or fact, is at first vague, meaningless, incoherent. Because of this, the scientist finds it necessary to set up or construct, vis-à-vis every given set of unorganized experiences, a model, originally invented by the human mind, which stands somehow in correspondence with the facts themselves. These constructs are not introduced arbitrarily without rhyme or reason. They are subjected by the scientist to certain requirements, called metaphysical requirements.[1] Among them are the following:

1. They must be fertile in a logical sense; that is to say, they must make a difference to what you are observing or explaining. An idea such as the Berkeleyan God who causes every event in the world by merely thinking about it, the theory according to which every happening in the world is a thought in the mind of God, provides no fertile scientific explanation because it cannot

[1] A more extensive account may be found in the author's *The Nature of Physical Reality* (New York: McGraw Hill, 1950).

be tested effectively; it does not furnish any way of understanding or of checking. All scientific constructs must satisfy the 'principle of logical fertility'; they must make a difference to your pursuits.

2. They must be extensible. Science rejects them if they merely illuminate a very small part of your immediate experience, only a few of the facts you wish to explain. Useful scientific ideas must have a fairly inclusive reach.

3. They must be richly interconnected among themselves, aside from their reference to facts. By virtue of this pervasive relationship, the ideas of science permit themselves to be controlled by general principles which are not directly suggested by the scientific enterprise itself. It is not allowable for a scientist to employ an idea in terms of which he can explain one or two observations while the idea does not set itself into some logical relation with anything else. Large internal coherence between constructs of explanation must be sought, although it is not always achieved.

4. There must be simplicity. A scientific theory must be formulated in terms, or in mathematical equations, which are in some sense simple. To be explicit, invariance with respect to transformations is the present version of simplicity in physical science.

5. There must be a certain elegance ruling the manipulations of the scientific constructs which serve as explanations of the obscure facts.

6. The principle of causality comes into play here too, but this is a rather technical philosophic matter which hardly needs to be discussed in the present context.

These, then, are the metaphysical requirements of every science. Notice that the list does not contain one which was held to be important and clinching in the sciences of the last century, namely, the requirement of detailed pictorability. The models used in modern science, the constructs of explanation, do not have to be couched in terms of *visual* concepts. We no longer explain the atom in terms of a group of electrons moving mechanically around a nucleus, each electron having a certain trajectory and occupying a definite point in space at a specified time. The electrons of today do not have paths any longer. Their motions are not visualizable at all, and this for various reasons like the following:

Electrons are far smaller than a wavelength of light and hence they have no colour. Their position cannot be ascertained by the usual kind of physical experiment, which involves the reflection of electro-magnetic or other kinds of signals from the entities themselves. Such procedures would fail simply because the entities are too small. Trying to find out where an electron is would be very much like trying to find out the position of a ping-pong ball by shooting cannon balls at it. No, in a very fundamental sense these entities have lost the facile attributes of localizability in space and time. We operate in terms of probabilities, and these probabilities may be the irreducible determinants of natural happenings. What I am saying here is that the requirements upon the constructs in terms of which we now explain the physical world have become exceedingly elusive and abstract.

So much for the *metaphysical* requirements. There is also the well-known *empirical* requirement of *verifiability*, of experimental or observational confirmation.

I sometimes find it useful to present the pursuits of science, its methodology, by means of a graphical analogue. Imagine with me all the obscure facts, all the unconnected observations, these contingencies which do not explain themselves, as being mapped on a certain plane, the 'protocol' plane of human experience. From that plane we can go by means of rules of correspondence into the domain of theory, where you manipulate logically fertile constructs. After passing into that domain, you can now calculate, you can now reason. Reasoning amounts to an ideal movement among, to a transformation of, the constructs of explanation. Such a procedure, which is essentially that performed by the theoretical scientist, leads to a certain place within the field of constructs where you can again use a rule of correspondence (operational definition) which permits you to return to the protocol plane of experience, to 'Nature' if you please. This picture, together with a more detailed exposition of the scientific method, can be found in the book already mentioned (note 1).

Now, that latter transition is a prediction. The empirical requirement of verification insists that a circuit from a certain point in the plane of protocol observations into the domain of theory and back again to observation shall be succesful. In other words, you must be able, as outlined elsewhere, to predict, not on the basis of some *a priori* kinds of concepts, but to predict on the basis of having previously injected into the scientific theory a material bit of knowledge at some other place in our diagram.

Forgive me for being so technical in my discourse. I shall soon

ESP IN THE FRAMEWORK OF MODERN SCIENCE

try to be much more specific and concrete, but I thought it well at the beginning to say a few things about the workings of scientific theories in general.

There is one further point of rather general import which I must not leave unmentioned. Old-style science, the physical science of the last century, believed that its fundamental premises were secure and not subject to question or to change. To be sure, scientists spoke of postulates, of axioms. These included the basic principles of arithmetic (the postulates concerning natural numbers), the basic principles of geometry (the postulates of Euclid), the basic principles of physical science (the law of conservation of energy, momentum, etc.). These were Truths, with a capital T, firmly embedded in human knowledge, truths which had to be accepted by every sane mind, truths which carried within themselves the affidavits of their validity. Now this attitude has changed. The modern scientist has learned that even his postulates are held on trial. The whole business of science is in a flux: science is a progressive, self-corrective, dynamic enterprise which subjects itself, in response to the ever-present threat of falsification, to repeated changes and revisions of its fundamental tenets. To put it bluntly, science no longer contains absolute truths.

We have begun to doubt such fundamental propositions as the principle of the conservation of energy, the principle of causality, and many other commitments which were held to be unshakeable and firm in the past. And this has, I think, an interesting bearing upon your own pursuits, for it means that the old distinction between the natural and the supernatural has become spurious. That distinction rested upon a dogmatism, a scientific dogmatism, which supposed that everything in the way of fundamental facts and basic matters was known and that there was an obvious distinction between what was possible and what was not possible. Today we know that there are many phenomena on the fringe, at the periphery of present-day science, which are not yet understood, which are still obscure, but which will nevertheless be encompassed by the scientific method and by scientific understanding in the future.

An analogy which does not appeal to me as a description of science is that of the picture puzzle, the pursuit of finding facts, then putting them together and somehow discovering a pattern in them, a pattern whose completion closes and solves the scientific problem. This analogy is incorrect because you cannot resolve a scientific problem in any manner that is ultimate.

I like to think of science as a crystal which grows within an amorphous matrix of liquid experience. This amorphous matrix—unorganized, interesting because of its caprice, the chicanery of chance which it embodies—is most important in our lives. But somehow our minds are disposed to organize this material, and the organization is very much like the growth of a crystal within its liquid environment. Nobody can tell where the crystal is going to grow. Starting unpredictably from some small seed, it stretches out a long arm in one direction, or it proceeds on a broad front along another. But whatever it does, it changes the amorphous liquid into a pattern, a lattice, which makes prediction possible, which confers order upon the constituents of the matrix and understanding upon those who perceive it. Now this process of crystal growth is not self-limiting—if the vessel permitted it, the crystal would grow for ever. And yet it would never exhaust all the infinite supply of liquid. The crystal of science will go on growing for ever. And since the number of facts, these immediate experiences in the protocol plane of which I spoke, is practically infinite, one sees no limit at all to the growth of science and there will probably never be a time at which all the liquid, unorganized and therefore occult matrix of human experience will have become crystallized into a rigid scientific pattern.

Now science contains a great number of unorganized facts, not only those of psychical research. There are many, many experiences surrounding us on all sides which are at least as challenging, at least as mysterious, as those upon which you have bestowed your attention. Let this not be forgotten. Here I am coming down, in somewhat more definite terms, to matters that stand for discussion. I would like to speak of three effects, clairvoyance, telepathy, and precognition, albeit very briefly and inadequately, for I lack a complete knowledge of the facts. These facts, incidentally, I shall assume to be correct, since I have no competence to doubt them. It would seem just as unreasonable for me to doubt what Dr. Murphy and other psychologists of high repute tell me about parapsychology as it would be to doubt the reports of highly esteemed astronomers like Dr. Allan Sandage about quasars. As a matter of fact, I am as disturbed in my own mind, and at the same time fascinated, by the things I read about quasars as I am by clairvoyance, telepathy, and precognition. Acquaintance with some of the active researchers in the field of psychical research has given me an extremely high regard for their ability to judge evidence and detect deception, an ability which I have at times found wanting among friends in the physical sciences where, because of greater stability of knowledge, it is no longer needed.

Now, in connection with clairvoyance, I think that first of all it must not be forgotten that all primary knowledge—the knowledge which is finally crystallized in the various sciences—arises from within consciousness. Consciousness is the primary medium of all reality. Even the external world is initially a posit, a projection of consciousness, which can be tested by its consistency with other items of consciousness, with the totality of human experience. That projection, the external world, takes on ontological existence after being tested and confirmed within conscious experience along the lines of scientific method which I outlined so vaguely a few moments ago. To believe in the 'existence of an external world' aside from its consistency with experience is an ontological commitment which we need not face in the present context. Let me merely say that if you stipulate the transcendental existence of an external world outside of mind, you are making an ontological leap—a natural one, for everybody makes it—which involves the freezing of a certain epistemology by fiat into a definite form.

At any rate, if we employ the methods I have outlined, we arrive at a theory of *ordinary* perception which I would like to illustrate by means of a very simple diagram (figure 1). Let me picture a person's consciousness (C) as a circle. Now, right round that circle is a concentric ring which I will call body (B). The body is in immediate contact with the mind, with that area of consciousness within which all experiences arise. Now let the circle to the right represent an object (O), a physical object. There are physical effects (PE), effects which by now are reasonably well understood and which somehow link the object with the layer around consciousness called the body. The study of the interactions between the object and the body falls under the principles of physics. Now the miracle in all this, at the present time, does not reside within this field of interaction between the object and the body; the miracle occurs when anything that takes place in the body emerges within consciousness. We don't know how this happens at all. We do have some rather vague theories—psychophysical parallelism, panconsciousness, and all the rest—but the point I am making is that the phenomenon of ordinary perception of the object out there, the ingression of the effects of that object into consciousness, poses an obstacle to understanding tantamount to a miracle. And that miracle is the conversion of the physiological stimulus into a conscious response.

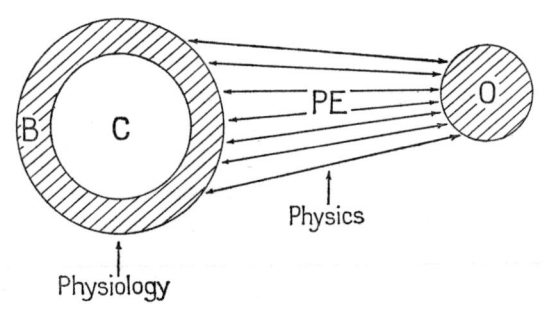

Figure 1. Schematic representation of the interactions between object and body in ordinary perception.

The relation between physical effects within the body and the emergence of experience within consciousness is *not* strictly a causal relation. There is no theory of cause and effect which can validly, fruitfully, contain this kind of transition. From the scientific point of view, it is just as mysterious as anything we encounter in the field of parapsychology. Science, as I have said, regulates the transmission from O, the object over there, to the periphery of consciousness, but it does not explain at present how that periphery is crossed.

Let me now recall to you briefly the laws of physics in their most general features for there may be a lesson in them for psychi-

cal researchers. In this transition from the object to the body we have discovered the following things:
First of all, there is the principle of conservation of energy. This is an all-pervasive law which is partly being read from nature, partly being injected into nature. It is not merely a generalization of empirical fact because whenever we find something which violates the principle, we are tempted to invent a new form of energy. Let this be acknowledged. Helmholtz 'proved' the principle of conservation of energy in the 1850s, but his proof involved the assumption that there exists a new kind of energy previously unknown, namely, potential energy. Only by introducing this as an additional concept was he able to prove the law. The addition was proper because the concept proved fruitful; it satisfied all the metaphysical requirements.

Today, the principle of conservation of energy is thought to hold in nearly all domains. However, it is not valid without exception. At the forefront of current physical research, in the fields of quantum theory and elementary particle physics, the principle of conservation of energy is frequently breached because we find it necessary to invoke the existence of 'virtual processes'. Virtual processes do not conserve energy. They follow no ordinary law, but are confined to extremely short durations. In a very short time, every physical process can proceed in ways which defy the laws of nature known today, always hiding itself under the cloak of the principle of uncertainty, to be sure. The point I am making is that when any physical process first starts, it sends out 'feelers' in all directions, feelers in which time may be reversed, normal rules are violated, and unexpected things may happen. These virtual processes then die out and after a certain time matters settle down again in obedience to the principle of conservation of energy. The term 'virtual' is not synonymous with 'unreal', for these processes cannot be ignored without falsifying the scientific prediction of actual events.

A great deal is made of the fact that physical processes decline in intensity with an increase in distance. I was greatly interested in Dr. Osis' remarks this morning concerning the possible relations between ESP and distance. This appears indeed as a worthwhile field of experimentation. It is quite true that most physical processes follow an attenuating law of some sort. However, it should also be recalled that not all interactions obey an inverse square law—in fact, almost none do. Only interactions between physical points follow an inverse square law, and strictly speaking there are no physical points. An electric field in front of a charged plane of infinite extent shows no attentuation at all. It would be as strong at a distant star as it is right in front of the plane. There are 'resonance forces' encompassed in modern quantum theory, which may possibly have interesting applications in biology and in psychology and which decrease very slowly with distance.

Today physicists hear a great deal about Mach's principle. This principle in effect says this: Newton's laws as they are usually understood are nonsense. Inertia is not intrinsic in the body at all; it is induced by the circumstance that the body is surrounded by the whole universe. It is for this reason that a force is needed to accelerate an object. Thus Mach believed, tentatively at least, that the existence of inertia in an isolated body is a consequence of the fact that there are distant masses of stars around it. We know of no physical effect conveying this action; very few people worry about a physical agency transmitting it. As far as I can see, Mach's principle is as mysterious as your unexplained psychic phenomena, and its formulation seems to me almost as obscure.

Men in theoretical physics today invoke a principle known as the 'exclusion principle'; it was discovered by Pauli. The exclusion principle is responsible for most of the organizing actions that occur in nature. We actually speak of 'co-operative effects'. All of these are brought about by the so-called Pauli principle, which is simply a principle of symmetry, a formal mathematical characteristic of the equations which in the end regulate phenomena in nature. Almost miraculously it calls into being what we call exchange forces, the forces which bind atoms into molecules and molecules into crystals. It is responsible for the fact that iron can be magnetized, that matter cannot be squeezed together into an arbitrarily small volume. The impenetrability of matter, its very stability, can be directly traced to the Pauli exclusion principle. Now, this principle has no dynamic aspect to it at all. It acts like a force although it is not a force. We cannot speak of it as doing anything by mechanical action. No, it is a very general and elusive thing; a mathematical symmetry imposed upon the basic equations of nature producing what appears like a dynamic effect.

Towards the end of the last century the view arose that all interactions involved material objects. This is no longer held to be true. We know now that there are fields which are wholly non-material. The quantum mechanical interactions of physical psi fields which play an important role in the theory of measurement —interestingly and perhaps amusingly, the physicist's psi (the square root of a probability) has a certain abstractness and vagueness of interpretation in common with the parapsychologist's psi —these interactions are wholly non-material, yet they are described by the most important and the most basic equations of present-day quantum mechanics. These equations say nothing about masses moving; they regulate the behaviour of very abstract fields, certainly in many cases non-material fields, often as tenuous as the square root of a probability.

Finally, there has emerged in these studies the view that there can be no instantaneous action at a distance, that there must be causality of a specific kind. This means that there can be no causal connection between two events at different points in space if they are farther apart than the distance which light could travel within the time interval between the two events. Such is one meaning of causality in modern physics. It imposes a real limitation upon what can actually happen in the world. Yet, if you assume the correctness of that principle (and there are few who doubt it), you will not get any effective limitations upon what you report as being the case in your studies of ESP and other paranormal occurrences. For the restrictions imposed by causality involve events so far apart in space or so close together in time that they would hardly come under your observation with the techniques presently available. I am saying that the principle of causality as it is now conceived by the physicist is almost without significance for paranormal effects.

The foregoing remarks were meant to show that physics, indeed all so-called exact sciences, are not closed books, that they contain many unresolved problems which are not entirely without analogy, and perhaps even of some relevance, for the worker in psychical research. They also suggest that current physics is different from the science of the last century and that some of its ideas are as difficult to grasp as those of parapsychology.

Now, let me speak very briefly about a few present possibilities which are sometimes invoked for explaining clairvoyance. Here, I fear, my conclusions will be somewhat discouraging. In quantum mechanics one meets an effect which is called the 'tunnel effect'. It implies that if you set up an obstacle between two bodies which would normally stand in physical communication via electro-magnetic signals, through the passage of photons, electrons, or other kinds of particles (the situation is perfectly general) you cannot with rigour exclude all possible transmission of effects, all communication. To be sure, in classical physics, if a particle coming from one body and going to another does not have enough energy to surmount the obstacle or barrier, it simply cannot get through. According to quantum mechanics, this is no longer true. If a lot of particles—a lot of photons, electrons, neutrons, or whatnot—are emitted by an object, some *will* get through under practically all conditions. In other words, complete screening is no longer possible. Now, I am not suggesting that you can explain clairvoyance by the transmission of a few photons from the distant event to the person who perceives it clairvoyantly. Nevertheless, there is perhaps something here to be thought of a little further. But I see no great promise in this direction.

Another thing which I cannot leave unmentioned is the possibility of a supra-conscious inference from present clues (this would apply to telepathy and precognition as well as to clairvoyance). I know this has often been thought of by psychical researchers, but it seems to me so important that perhaps it should be reconsidered. I don't quite know how it happens, but our minds seem to be endowed at times with a peculiar clarity in which we see and understand clues present in a given situation in a way which seems wholly incomprehensible when viewed as commonplace perception. I myself had this sort of experience when I was young. We played a game and we called it telepathy. Of course it had nothing to do with strict telepathy, but its full explanation still eludes me. Later we called this parlour game muscle reading; undoubtedly you are familiar with it.

One person went out of the room and the rest of the group decided on something he was to do, some action he was to carry out such as going from one room to another, opening a drawer, taking out a certain object, taking it to another room, finding a person, putting the object in his pocket, taking something out from another pocket, and so on—a long series of tasks like that. Then, when this had been decided upon, the subject would come back into the room and a member of the crowd would grasp his left wrist and think very hard, very intensely and seriatim, about the actions the subject was supposed to perform. And believe it or not, he went right ahead and carried them out correctly. Now, this is a very simple thing. Most people could learn to do it in a few days and the success was very striking to the onlooker.

While this procedure seems miraculous, I am sure it is capable of being understood and analysed in terms of ordinary physiology and psychology. The subject adjusts himself very delicately to the impulses he gets from the other person's hand and proceeds to carry out the required actions. There were those who could perform similar feats even without bodily contact, chiefly, I believe, by scrutinizing the facial expressions and the involuntary gestures of the onlookers. I'm not sure that something like this, supraconscious inference from present clues, may not have something to do with clairvoyance and some of the other phenomena in which you are interested.

Now let us leave clairvoyance and turn to telepathy, the transfer of information from mind to mind. Look again at the diagram illustrating ordinary perception (figure 1). I want to change the circle marked O (for object) on the right to represent another person. There will then be the area of consciousness (C) surrounded by the concentric ring which I called body (B). Now the ordinary explanation of communication, which involves first a conversion of a conscious stimulus into a bodily response and calls upon the laws of physics to explain communication of the signal from one body to another, and then finally the conversion of a physiological stimulus into a conscious response, contains *two* mysteries. Let this be carefully noted. Now, if one is willing to accept the possibility of telepathy, of direct contact between minds, he thereby reduces these two mysteries to one. And speculations concerning such a possibility are in my opinion by no means unscientific *per se*; as a matter of fact, from the point of view of simplicity such a thesis is more satisfactory than the view which involves two mysteries. You know, of course, that many forms of religion are based upon the possibility of direct, bodiless contact between God's consciousness and man's.

I have already mentioned muscle reading in connection with the unravelling of semi-conscious clues. One might properly wonder whether, in the absence of a scientific explanation at the present time of ESP and related phenomena, a little more thought should not be given to those occurrences which lie on the boundary between the normal and the paranormal—to 'semi-normal' phenomena. I cannot believe that there are any discrete, quantum-like transitions in human experience. There must be an area where wholly inexplicable things merge with those that we understand as normal happenings. Let me illustrate:

When I was a youngster of fourteen, I had to take an examination in a distant town. I had to go away from home for the first time. The examination was a three-day ordeal. On the evening before the last day I tried to recall what I had learned about geology because I knew perfectly well that we were going to be quizzed the next morning on the geologic periods. To my great frustration, I couldn't recall them at all; I knew nothing about them. I went to bed in great perturbation. During the night I had a dream: I found myself at home, took my geology book from the shelf, opened it, and read the page on the geologic periods. I knew them the next morning, and I passed my examination.

Now I don't know whether this is anything of interest to psychical researchers, but I do think it is one of these semi-normal boundary occurrences. And perhaps an examination of the things which we half understand and therefore ignore might lead towards an opening of comprehension of those phenomena which are at present completely obscure.

A few words remain to be said about precognition, a field which seems rife with conjectures of a quasi-physical sort. Unfortunately, I know of no physical theory available at present which can be drawn upon to explain temporarily prior knowledge of coming events—except again through an analysis, perhaps unconscious or supraconscious, of pre-existing causal clues. Two hypotheses have been invoked most frequently to account for inverted knowledge of a causal sequence. One is Feynman's theory of time reversal. It is alleged to permit time to flow backwards, since it assigns meaning to trajectories of electrons in which the time-axis is reversed. These are, in fact, normal paths of antiparticles, positrons in this instance, on which the latter travel forward in time.

The time concept has been greatly distorted by philosophers who fail to distinguish between time as a conscious, protocol experience, the 'stream vector of consciousness', and time as a theoretical construct. No physical theory is qualified—by virtue of the methodology outlined earlier—to say anything about the structure of subjective time. Physics deals with measured, objective time, which means the construct. This, however, is connected with the stream of consciousness by rules of correspondence which must conform to immediate experience. To put the matter simply, even if constructed time flows backwards, as in the case of the normal motion of a positron, the relation between constructed time and consciously experienced time *must* be such, and is meant to be such, as to leave conscious time flowing forward. I have dealt with this and other related matters elsewhere.[2] This does not rule out such radical reinterpretations as precognition might require. The point is that contemporary physics provides no example of it.

Another instance I have seen cited focuses on the circumstance that quantum electrodynamics features Feynman diagrams in which the effect precedes the cause. A nucleus is supposed to explode before the missile reaches it. But the correct interpretation of these diagrams recognizes no causal connection in these cases: the nucleus *happens* to explode with a certain (very small) probability spontaneously before the missile arrives, and a causal sequence is simulated by the fact that the emitted positron *later* collides with an electron passing near the nucleus in an act of mutual annihilation.

An artifact occasionally invoked to explain precognition is to make time multidimensional. This allows a genuine backward passage of time, which might permit positive intervals in one time direction to become negative ('effect before cause') in another. In principle, this represents a valid scheme, and I know of no criticism that will rule it out as a scientific procedure. If it is to be acceptable, however, a completely new metric of space-time needs to be developed, a metric which will account not only for the facts to be explained, but also for the known laws of physics which can be fully understood without it. Remember the principle of extensibility!

I have probed physics for suggestions it can offer towards a solution of the sort of problems you seem to encounter. The positive results, I fear, are meagre and disappointing, though perhaps worth inspection. But why, I should now like to ask, is it

[2] Henry Margenau, 'Can Time Flow Backwards?' *Philosophy of Science*, Vol. 21, 1954, pp. 79–92.

necessary to import into any new discipline all the approved concepts of an older science in its contemporary stage of development? Physics did not adhere slavishly to the Greek rationalistic formulations that preceded it; it was forced to create its own specific constructs, even to the point of denying the basic geometry of Euclid. Lo and behold, these were later shown to be compatible with, and often generalizations of, earlier more primitive notions.

Hence I should think that the parapsychologist need not be discouraged by the lack of suitability of physical ideas, or ideas strictly based on present-day physics, for his purposes. If his facts are clear, reproducible, and beyond the vagaries of chance, then I see no reason why he should heed the objections of unimaginative colleagues in the physical sciences.

The parapsychologist, I think, is not likely to find theories which will illuminate his area of interest already prepared by physicists. He must strike out on his own and probably reason in bolder terms than present-day physics suggests. The only bridle upon his speculations is occasioned by the need for empirical verification and by those clearly recognizable metaphysical principles which control all of science. The concepts of parapsychology may well turn out to be at first completely different from the concepts of contemporary physics. The other behavioural sciences are not fashioned precisely after the patterns of inorganic behaviour; yet they are acceptable and they succeed. And so, if I may return to my initial story, I would say, 'For the present, let the darn telephone ring!' Tolerate the strident critical voices of hard-boiled, pragmatic, and satisfied scientists without too much concern, and continue his own painstaking search for an understanding of new kinds of experience, possibly in terms of concepts which now appear strange.

X

THE FEASIBILITY OF A PHYSICAL THEORY OF ESP

Adrian Dobbs

By a 'physical' theory I mean one which involves only agencies and laws of the kind acceptable in principle to physics as currently practised. Specifically, in a 'physical' theory in my sense, it is permissible to introduce only agencies located in physical space-time, which interact according to laws of the kind recognized in classical physics and chemistry, and latterly in relativity theory and quantum theory. Applied to the processes of perception (normal and paranormal) and the other functions of a human being, such a theory would be an instance of what Professor Price has called a 'Materialistic' or 'Naturalistic' theory of human nature ('Psychical Research and Human Personality').[1] Price calls any theory 'Materialistic' (or 'Naturalistic') which maintains 'that mental processes of every kind are unilaterally dependent upon physico-chemical processes in the brain. . . . This, or something like this, is the Materialist conception of human personality, technically called Epiphenomenalism.'

Price admits that this theory is a most impressive one when the empirical facts of normal sense-perception are considered. But he argues that it is not compatible with at least one kind of paranormal perceptual phenomenon when he says, 'I think that if we consider the implications of Telepathy, the most elementary and the best established phenomenon in the whole field of Psychical Research, we shall see that they are incompatible with the Materialistic conception of human personality.'

[1] See Ch. 3 in this volume.

I want to consider Price's argument in some detail because I think it is a particularly clear and cogent expression of a point of view which is responsible for a good deal of the current prejudice on the part of natural scientists against parapsychological research. I believe that Price's point of view can be shown to rest upon questionable assumptions, which need to be weighed critically before the possibility of a physical account of telepathy and other forms of ESP is rejected. I shall therefore consider a number of passages in Price's article which purport to show that a physical explanation is out of the question. In doing so, I do not wish to be understood as maintaining a Materialist or Epiphenomenalist position in exactly the terms which Price has defined it. For one thing, I do not wish to endorse any view which stresses the old-fashioned ontological primacy of the concept of matter, since the basic concept of current physics is energy rather than matter in the sense of mass. This change of viewpoint entered physics with Einstein's demonstration in 1905 of the equivalence of mass with rest-energy in the Special Theory of Relativity. Consequently I should not wish to accept the Materialist label (with any nineteenth-century implications) for the kind of view which I am seeking to put forward and which I should prefer to call 'physicalist'. Indeed, I think that the point of view I am adopting is just the one now adopted by most scientists, which has been well summarized by one of the world's leading physiologists, Sir John Eccles, who would not normally I believe be regarded as a 'Materialist'.

In an article entitled 'The Physiology of Imagination',[2] which is devoted to the creative mental processes in all fields, he wrote as follows:

> Each of the preceding articles in this issue of *Scientific American* is concerned in its way with a process that goes on in the sheet of grey matter ·1 inch thick and 400 square inches in area, which forms the deeply folded surface of the two great hemispheres of the brain. This statement contains a premise that is best made explicit. It says all mental activity, including the supreme activity of creative imagination, arises somehow from the activity of the brain. Few would deny this premise, though a wealth of philosophical disputation lies concealed in that non-commital word 'somehow'.

[2] Published in a special issue of *Scientific American*, Vol. 199, 1958, pp. 135 ff.

THE FEASIBILITY OF A PHYSICAL THEORY OF ESP

I am merely concerned to argue against Price's claim that we cannot account for ESP in terms of entities and processes that are physical in the sense I have described; and to suggest possible ways in which we may hope to assimilate ESP to the kind of processes which physicists and physiologists so successfully invoke to explain normal sense-perception.

The main argument in Price's paper which I wish to contest is expressed in the following passage:

> In telepathy one mind affects another without any discoverable physical intermediary and regardless of the spatial distance between their respective bodies. The Materialist once we can get him to admit the facts, will no doubt try to explain them by physical radiations of some kind. Indeed that is what he *must* do if his conception of personality is to stand. But no explanation of that kind seems to be feasible. Such physical radiations, if they exist, ought to be detectable by physical instruments. It ought to be possible to intercept them *en route*; and their intensity should vary in some way with the spatial distance between the body of the agent and the body of the percipient. But none of these consequences, which ought to follow if the Radiation Theory is true, is in fact verified. If the supposed physical radiations do not have any of the empirically verifiable properties which physical radiations have, what is the point of calling them radiations at all, and what is the point of calling them physical either? They cannot be physical in the ordinary sense of the word 'physical', the one which the Materialist is using when he says that all mental events have physical causes. For in the ordinary sense of the word 'physical' nothing is a physical event or entity unless it is perceptible by means of the sense organs; either directly, or indirectly by means of instruments which can themselves be directly perceived.

It will be seen from this passage that Price argues that in telepathy one mind affects another without any physical agent or process as intermediary, because he *assumes* the following to be true of all agencies or processes which can properly be called 'physical' in any reasonable sense.

(i) They must at least be detectable by physical instruments, even if they are not perceptible through the senses.

(ii) Such detection must be able to reveal these influences *en route* between the telepathic agent and the telepathic percipient.

(iii) The intensity of such physical influences ought to vary in some way with the distance in physical space between the body of the agent and the body of the percipient.

Price further seems to assume that no influence which meets these conditions has in fact been found; and he concludes that this disposes of the physicalist or materialist position, and of 'the hypothetical physical radiations which are alleged to be the cause of telepathy. They have just been postulated *ad hoc* because no physical explanation of telepathy is possible in the ordinary sense of the word "physical"'.[3]

I believe that this conclusion is not warranted by such established facts as we have about telepathy, if we are using the word 'physical' to mean an agency or process of the kind recognized in principle in physics today. In what follows I shall contest each of Price's grounds that I have distinguished above for saying that no physical explanation of telepathy is possible. For reasons of convenience in the exposition of my argument I shall start by taking first the third of Price's contentions above: that if telepathy were due to a physical influence the intensity of such influence should vary in some way with spatial distance. My reason for doing so is that I believe that Price's position here is based upon a failure to analyse the implications of the facts, due to hidden assumptions which vitiate at the outset most discussions on this question. These hidden assumptions are, I think, both basic and invalid; so that they must be brought out into the open before any useful discussion can proceed.

The first hidden assumption is that, if an effect is due to a physical influence, it must vary with distance either because of the universal dissipative forces of friction and resistance which generate irrecoverable waste heat in ordinary conductors; or on account

[3] Price assumes that his 'Materialist' *must* try to explain by *'radiations'* of some kind'. I do not know exactly what he means by this term 'radiation' which is according to common usage employed to describe any form of emission of a physical influence from a source. There is certainly no justification for foisting in advance upon the hypothetical 'materialist' any idea that such emissions must be of any particular physical kind of rays (such as those of light or heat); or that they must *diffuse uniformly* into three-dimensional physical space, rather than flow in a current along a conductor of some sort between the brains of agent and percipient. This is a point of considerable importance from the point of view of the possibility of a physical explanation of ESP, and I shall have more to say later about the implications of the facts which have now been established by physics in the field of electrical superconductivity and the 'ducting' of radio waves.

of the well-known inverse square law by which electro-magnetic and gravitational effects are diminished in three-dimensional space. But the first part of this assumption, which was formerly held by nearly all physicists, has in recent years received striking refutation by the well-attested phenomena of electrical superconductivity. It is now known that in many metals, when cooled to the temperature of liquid helium, an electric current can be made to flow without any loss due to resistance and irrespective of the distance for which it travels. In fact, the only thing which attenuates the current is an unpredictable fluctuation in the molecular conditions of the conductor. Apart from such fluctuations a superconducting metal will allow the current to flow indefinitely; and there is now some prospect of designing materials which will be superconducting at room temperature.[4] As to the second part of Price's first hidden assumption, regarding physical radiations into three-dimensional space, I shall shortly show that the inverse square law does not necessarily apply, even to ordinary electromagnetic radiations. For in the case of radio waves the strength of a signal as received may even increase with increase of distance between transmitter and receiver, due to ionospheric effects and various forms of 'ducting'.

The second hidden assumption is that having regard to cases of telepathy reported over large distances, it must be a process of some sort of direct interaction in which one mind affects another 'regardless of the spatial distance between their respective bodies', so that the intensity or effectiveness of the response is not diminished by distance. I shall argue that this hidden assumption, that the intensity of the effect is unaffected by distance, is invalid on the basis of the recorded facts.

But first what are the facts? We certainly have a number of well-attested claims that telepathy has occurred over very large distances as well as over short distances. Many of these claims are to be found in reports of the so-called 'spontaneous cases' and some among the experimental cases. A number of spontaneous and early experimental cases are discussed in Chapters IV and VI of F. W. H. Myers' *Human Personality and its Survival of Bodily Death*.[5] Myers seems to have been particularly struck by the

[4] D. L. Atherton, 'Electrostatic Switching of One-dimensional Superconductors', *Nature*, Vol. 205, 1965, p. 687.
[5] Abridged edition (London: Heffer, 1913).

records of experimental induction of hypnosis by telepathy achieved by Dr. Gilbert at Le Havre. The distance between Dr. Gilbert and his subject, Madame B, was over 1 mile. Much more spectacular experimental results of a similar nature are claimed in Professor Vasiliev's book *Experiments in Mental Suggestion*.[6] For example, in Table 17 on p. 105, Vasiliev records results indicating that telepathically induced hypnosis took effect as quickly over a distance of 1,700 kilometres (Sebastopol to Leningrad) as over a distance of 25 metres. The conclusion which Vasiliev himself drew from these results is contained in the following passage:

> Analysing the results of our experiments at various distances, we find that these results are almost identical with those obtained at short distances. . . . Experiments, however, provide reasons for supposing that a similar mental effort is required from the sender in long as in short distance experiments.[7]

The argument, implicit in Price and explicit in Vasiliev, is that equally good telepathic results are obtainable over great distances as over small distances, *with the same attendant background conditions in both cases*. This latter clause is essential to the assumption to be found in both Price's paper and Vasiliev's book: that great variation in the distance between the bodies of agent and percipient has not affected *per se* the degree of success of the telepathic process, nor the intensity of its operation and results. There are two distinct points here.

First, accepting that telepathy has occurred over great distances, as well as over small distances, it does not follow that distance is wholly irrelevant to its occurrence, or *to the probability of its happening*. We have, for instance, no systematically compiled data to test whether it has happened as frequently over long distances as over short distances, taking into account the number of occasions when it has been tried experimentally. We have no relative frequency tables showing that percentage of success in all cases is independent of distance, so far as I am aware. So how can we tell whether in fact telepathy is in any different position to short-wave radio as a means of communication, as far as relative insensitivity to distance goes? Every experienced operator of radio transmitters knows that 'breakthrough' conditions occur sporadically when signals are picked up 'loud and clear' over dis-

[6] English translation, Inst. Mental Images, 1963.   [7] *loc. cit.*, pp. 105-7.

tances far in excess of those their transmitters are designed to reach under normal working conditions. This is the phenomenon of 'ducting' well known to radio physicists and engineers, and attributed to fluctuations in portions of the Earth's atmosphere which reflect radio waves, and which in certain conditions provide a specially favourable path to certain selected frequencies for limited periods of time. In these conditions the favoured frequencies are conveyed along the ducting path with very much less than the usual dissipation of energy.

But apart from these exceptional 'ducting' conditions there are the *normal* facts of short-wave radio transmissions. The first person, so far as I know, to point out the striking nature of such facts, about the apparent insensitivity of ordinary short-wave electromagnetic radiation to distance, in the context of telepathy, was G. W. Fisk, doyen both of amateur radio transmission and of experimental psychical research. Fisk reported in a communication to the *S.P.R. Journal*,[8] January 1935, the following remarkable phenomena. He had a 50-watt transmitter located at Tientsin in China on which he broadcast test signals on a wavelength of 30 metres (10 megacycles frequency) with the following results, recorded by different listeners at the locations and distances shown in the following table:

Location	Distance in miles	Audible strength of signal received ($r = 1$ barely audible; $r = 9$ very loud)
Chefoo	180	Nil
Shanghai	800	$r6$
New South Wales	6,000	$r3$
New Zealand	7,500	$r5$
Argentine	15,000	$r7$

Thus the loudest signal of all was in fact heard at the greatest distance of 15,000 miles; and the next loudest signal at a distance of only 800 miles; while at the closest station (which happened to lie within the 'skip' distance) no signal at all was heard. Even more striking results [9] were obtained in experiments across the Pacific with another enthusiastic amateur transmitter—a Colonel Foster

[8] Vol. XXIX, p. 35.
[9] Fisk did not communicate these results to the S.P.R. for publication at the time, but he has communicated them to me in a personal letter from which I am permitted to quote.

in California—with whose help Fisk tried to see what was the weakest signal power which could carry an intelligible message across the ocean. This was found to be between one and two watts (the power of a pocket torch bulb). Of course, this is still immensely more powerful than the level of electro-magnetic power found by the electro-encephalograph on the surface of a human scalp, which is of the order of 10^{-18} watts. This level of power, and the very low frequency available for electro-magnetic radiation from the human brain, makes it exceedingly improbable as a medium for telepathy. But the reason for rejecting brain radio as an explanation of telepathy is not as given by Price: the insensitivity of the telepathic effect to distance. The real reason is simply the sheer weakness of the source of power and its very low frequency, corresponding to immensely long waves. Price's contention so far by no means rules out short-wave electro-magnetic radiation of high frequency because he has neither adduced the relevant known facts about the reception of weak radio signals at great distances; nor the known facts about the electrical activity of the human brain. The point I am making here is that one cannot throw overboard the possibility of some physical basis for telepathy on the score of its insensitivity to distance, without discussing the apparent insensitivity to distance of short-wave radio reception. Such discussion also makes it clear that propagation conditions and their effects on the channel of communication connecting transmitter and receiver must also be taken into account.

The second point which arises in connection with the 'insensitivity to distance' argument against a physical basis for telepathy turns on the idea that, as the late G. W. N. Tyrrell contended, if there were, in fact, such a physical basis 'it is pretty safe to say we should get an unmistakable gradation at short distances if the effect were physical. When agent and percipient were sitting with their heads a few inches apart there would be a far more striking result than when they were in adjoining rooms.'[10] This is an argument which Tyrrell (himself an electrical engineer) used to try to dispute Fisk's contention that the inverse square law of variation of the intensity of an electro-magnetic wave with distance was irrelevant to the telepathy question. Tyrrell maintained that the special conditions of reflection at the ionosphere of

[10] S.P.R. Journal, Vol. XXIX, p. 41.

THE FEASIBILITY OF A PHYSICAL THEORY OF ESP

short radio waves, which gave them their relative insensitivity to distance, did not apply to measurements of field strength at short distances; and that at distances within the 'ground wave' as opposed to the 'sky wave' the inverse square law applied. Therefore one should compare the results of radio reception at short distances to get a proper comparison. This may be true enough of the particular case of electro-magnetic and other disturbances subject to the inverse square law of variation of intensity with distance; but how can one make such a comparison at all without introducing some notion of 'intensity', analogous to electro-magnetic 'field strength', into the notion of telepathy? This is indeed an essential preliminary for any scientific discussion of a hypothesis for telepathy, no matter whether the hypothesis be one of a physical or mental nature.

It is a curious fact that there has been practically no discussion of this problem of providing a telepathic 'field strength meter'. But without having defined some objective measure of the intensity of the effect, which results from the telepathic interaction between two minds, it is not logically appropriate to speak as Tyrrell does of expecting 'more striking results' the closer agent and percipient are together in physical space (on the assumption that telepathy is due to a physical cause). Nor is it possible to decide rationally whether there is evidence for or against the idea that the telepathic effect is wholly independent of the spatial distance between the bodies of the agent and percipient.

To take an extremely crude example. You might have a situation in which a thought impulse from agent A, that had been tremendously weakened by travelling 1,000 miles, was none the less just strong enough to be detected by the unconscious physical perceptual mechanism of a percipient C, so as to give rise to a physical impulse in his nervous system, and thus secondarily to trigger a pre-existing complex of strong emotional tone, finally giving rise to intensely vivid introspectible imagery. The resultant conscious mental state of the percipient due to agent A might then be one of *stronger* 'force and liveliness' than another conscious mental state in the same percipient, occasioned by the reception of a *stronger* thought impulse from a closer agent B which triggered a more weakly toned unconscious mental disposition. The content of these two distinct introspectibly conscious states in the mind of the percipient C might be such that we should say in both cases that there had been telepathy. But how could one decide which telepathic interaction had been the more intense, or say whether the telepathic field strength of A was stronger or weaker than that of B? One possible criterion which might be applied in *experimental* cases (where the agent's target thought is specially chosen to facilitate evaluation of faithfulness of reproduction by the percipient) would be the probability of the degree of success achieved having occurred by chance coincidence. This is in fact the criterion which has been applied in the so-called 'statistical' experimental work in psychical research; and it enables one to label results as 'more striking' the smaller their probability of arising by chance. It is rather similar to the criterion which is used in modern communications engineering to evaluate the rate of transmission of information in a *physical* system—a suggestive point.

I shall shortly have more to say about this criterion and the difficulties of applying it from the standpoint of physical communication (information) theory even in the simplest experimental cases dealt with by psychical research. At this stage I want to insist that, even if this criterion can be applied in such cases, it still does not give the complete answer to the basic problem of finding a measure of 'telepathic field strength', in the sense required to enable us to decide in any given case, whether a telepathic interaction was or was not insensitive to distance. For modern radio technology has shown that it is practicable for a receiver to detect exceedingly weak electro-magnetic signals; and by using systems of Automatic Gain Control, to amplify incoming signals, which are of very different strengths at the aerial stage, in such a way that both strong and weak signals appear at the output stage of the loudspeaker with subjectively equal audible strengths. Thus we are enabled consciously to hear the signals we are interested in at a more or less constant level. Consequently the *audible* level of signal strength at the output stage of a radio receiver is no test of the strength of the incoming signal at the input or aerial stage; which is the relevant factor when we are considering the effect of distance between receiver and transmitter upon the carrier of the information, both in radio and telepathy. For clearly, if some system of Automatic Gain Control were to operate in telepathy, it could offset attenuation of the signal produced by distance.

THE FEASIBILITY OF A PHYSICAL THEORY OF ESP

We must now take the argument a step farther and ask what is the measure of intensity of effect, or result, at the output stage of the receiver or percipient, in telepathy. Using the concepts and methods at present employed in psychical research by Price and others are we in a position to recognize and express the differences between 'strong' and 'weak' telepathic signals at the 'output stage' of our introspective consciousness, by some criteria analogous to 'degrees of audibility' quoted in Fisk's table above? Unless a satisfactory measure of effectiveness or intensity can be devised in terms of the resultant output at the stage of the percipient's conscious response, it does not seem possible to decide the question whether telepathy is independent of spatial distance or not. In the term 'conscious response' I mean to include both the 'behavioural' response characteristic of the Soal–Shackleton experiments and the 'introspective' reports given by percipients as descriptions of their own mental states in spontaneous instances of telepathy. From the introspective descriptions in the literature, one would expect the interaction of percipient C with agent A to be labelled more 'intense' than the interaction with agent B simply because a stronger emotional reaction had been generated in the percipient's introspectible consciousness in regard to A than B. I have already pointed out that a difference of this kind could be due to some idiosyncrasy in C, and thus to use it as a measure of relative intensity would be to allow the criterion to depend solely on the conditions at the input stage of *reception* and not of *transmission*; and we know that from the standpoint of physical communication theory this would be a distorted and misleading view. The method which physics has shown to be the right one for physical communication systems generally is to start by analysing all situations in which information is exchanged by physical means into five elements as follows:

Information [1] — input to transmitter [2] — channel [3] —
 Receiver of information [4] — Output of receiver [5]

Figure 2

This scheme shows no less than five distinct stages as numbered by indices above. The situation is then capable of precise quantitative analysis in two different ways. Firstly, it is possible to quantify the information at the input (stage 1) to the transmitter in a definite way, in terms of the system of symbols used to encode the information, with their respective *a priori* probabilities of occurrence, and to use a function called the 'entropy' of that information source. If the symbol probabilities are p_i then this quantity called Entropy $= -\sum_0^i p_i \log p_i$. Similarly, when the information is delivered to the final output of the receiver (stage 5 above) it is possible to measure its entropy or information content at this final output stage. One can then say how much of the transmitted information has been received, and how much has been lost in transit. Secondly, there is another and simpler method of estimating the losses in a physical communication system, namely by the direct measurement of the strength or power of the signal as it leaves the transmitter aerial, and the direct measurement of the strength of the signal as received in the aerial of the receiver. The ratio of these two quantities expressed in decibels gives the loss in *physical power* (as distinct from information) in transit. Provided the loss in power is not too great, the signal can then be amplified to any desired level necessary to recover the full information content.

The basic difficulty in the case of telepathy is that of finding a suitable measure either of information content or power to apply to the peculiar conditions in which information is ostensibly carried between two embodied minds, located apart in space, without any of the normal intervening channels of communication or peripheral sense organs. There is certainly no measure of *psychological* power of transmission analogous to the notion of power in the transmission systems studied by physics, where we can calibrate (within close limits) the power of the signal at the radio transmitter in watts; and measure the field strength in a receiving aerial at any given distance from it; and then calculate the attenuation or loss in power in transit in terms of decibels. So we are denied, in the case of telepathy, the simplest way of estimating losses of power due to distance available to radio engineers. But if telepathy is a genuine phenomenon it must be a form of communication of information (even in Price's view of 'direct interaction' between minds). But in order to establish that a particular case is a genuine communication of information, and not chance correlation of features of the content of two states of two differently embodied minds, we must have some standard or measure, how-

THE FEASIBILITY OF A PHYSICAL THEORY OF ESP

ever crude, by which to compare the information content of thought impulses in the agent with those of the percipient.

Vasiliev has introduced a useful new term 'telepatheme' for the information content of a thought impulse capable of being conveyed telepathically. In the cases where the telepatheme is the visual information of a picture, or a drawing, we can assign a more or less arbitrary measure to it, and we can see how much correct visual information is contained in the drawing made by the percipient. We can thus in principle measure the loss or attenuation of information in the telepathic transmission. This is the kind of system which is used to measure the efficiency of a television system where a precise evaluation of picture quality is required. It works in television because the picture is transmitted by means of a definite sequence of energy pulses each of which has to be separately received. It produces a precise quantitative measure of loss of information or distortion which is indispensable for the physicist, and could in principle be used to measure the information content or the loss of information in the telepatheme between the sending and receiving process.

But there are serious practical difficulties. Firstly one cannot see any way to measure the loss of information associated with the process of mentalization of the object (card, picture, etc.) which is to be telepathically transmitted. This process corresponds to stage 1 in the information diagram in figure 2 on page 235). One might guess that in the case of good visualizers there would be relatively little degradation of the information in the picture in the process of converting it into a telepatheme for telepathic transmission. But what about those who hardly visualize at all and tend to think of visual data in terms of sound images of the words describing characteristic features of those data? Here we would have to try to compare the information contents of a percept with that of a concept. This would be difficult though not logically impossible if we can imagine a method for enumerating all the characteristic features of the visual input data, and then conceive of examining the non-visualizer's conceptual description to see how many of these features fit the ones he describes. A marking system might perhaps then enable a crude assessment to be made of the loss of information involved, but it would probably be very inaccurate.

If we turn to the typical quantitative type of experimental

situation, that of guessing cards out of special packs of five kinds, it seems to be generally supposed that the only relevant information content is the decision which one of the five cards the target is. If this is the case, then there would not necessarily be any loss of information in the agent's converting his sense-datum of a card into a telepatheme which he wishes to convey to the percipient. In such circumstances the problem of assessing the degree of success, or effectiveness, of the telepathic interaction reduces to the purely mathematical problem of calculating the *a priori* probability of the percipient's getting a number of correct guesses by chance assuming satisfactory randomization of the packs. This is the usual procedure for evaluating the results of card-guessing telepathy experiments. But it rests on the assumption that, by merely looking at or 'fixating' a particular card picture, the agent is able to convert this into a 'telepatheme' containing all the relevant information content for unambiguous reception by the percipient, sufficient to enable his output behavioural response to specify correctly which of the alternatives it is. But we have no means at present of measuring the information content of the agent's *telepatheme*, as distinct from the information content of the agent's *sense-datum* of the target card. Vasiliev seems to suppose that the agent's sending or transmitting capacity to produce an effective telepatheme of maximum information content is proportional to the subjectively felt effort of concentration, so that the greater the felt effort the higher the efficiency of information conversion. But this again is pure assumption, though there does seem to be evidence for the idea that a necessary condition for telepathic transmission is the occurrence of at least a flicker of interest or slight surge of emotion on the part of the agent. But even if we try to equate such a surge of emotion with the power of the carrier wave of a radio transmitter we still have no counterpart in telepathy to the measure in radio of the depth of modulation of the carrier of the signal.

Turning now to the percipient or receiving end of the telepathic communication process we find a second but similar type of difficulty. For how can we measure the information content of the *telepatheme* as first received, when the only directly observable data we have are either the percipient's introspective reports of his *conscious* states, or the overt results of *decisions* made by him as a *result* of his supposed reception of the telepatheme? The decisions

I have in mind are those where, as in the Soal–Shackleton experiments the percipient does not report an introspectible image of the target card, but merely makes the behavioural response of writing down the initial letter designating its name. Just as there may be loss of information in the process of converting the input of sense data or images into the telepatheme for transmission, so there may be loss of information in converting the information in the telepatheme, as first received, into the conscious decision information, needed to cause the motor processes required to write a particular letter, or make a particular drawing.

It seems to me that here some other observations in Price's paper are highly relevant, though they do not in my opinion support his anti-Materialist (or, as I should say, anti-Physicalist) view, but rather the reverse. Price remarks:

> It looks as if telepathically received impressions have some difficulty in crossing the threshold and manifesting themselves in consciousness. There seems to be some barrier or repressive mechanism which tends to shut them out from consciousness, a barrier which is rather difficult to pass, and they make use of all sorts of devices for overcoming it. Sometimes they make use of the muscular mechanisms of the body, and emerge in the form of automatic speech or writing. Sometimes they emerge in the form of dreams, sometimes as visual or auditory hallucinations. And often they can only emerge in a distorted and symbolic form (as other unconscious mental contents do). It is a plausible guess that many of our everyday thoughts and emotions are telepathic or partly telepathic in origin, but are not recognized to be so because they are so much distorted and mixed with other mental contents in crossing the threshold of consciousness.

This is a very interesting and suggestive passage. It evokes the picture of either the mind or the brain as containing an assemblage of selective filters, designed to cut out unwanted signals on neighbouring frequencies, some of which get through in a distorted form, just as in ordinary radio reception. I agree with all of this; but I do not think it supports the case against the possibility of a Materialist or Physicalist theory of telepathy. To me, indeed, the supposition that the initial processes involved in the reception of a telepatheme are unconscious, and that a great deal of selective filtering goes on at an unconscious level, is a necessary ingredient in a physicalist theory. Price himself remarks:

The brain is a very complicated structure indeed. All kinds of other processes go on in it besides the ones which (in the Materialist view) produce conscious mental states; and some of these other physical processes might have unconscious mental states as their by-products.

These remarks are, I believe, very relevant to the next points I am going to take up in Price's critique of the Materialist–Physicalist approach. Before doing so let me summarize the broad conclusions we have reached, on the basic question of what grounds there are for holding that, if telepathic interaction is physical, this would be incompatible with the known facts about telepathy. First, we have seen that there are physical influences whose intensity is not diminished by the distance over which they have travelled, e.g. the current in a super-conductor; and secondly that unless we are given some facts about the information content of the telepathemes actually exchanged between agent and percipient, no valid conclusions can be drawn about whether the 'intensity' of the telepathic interaction is in fact affected by distance or not. We have found that it will not be sufficient merely to measure the information content of the target object or situation as initial input to the *agent*, and to compare this with the information content of the *percipient's* output response, either as recorded in his introspective report or as revealed in the outcome of his behavioural decisions. For such a procedure makes no allowance for information lost in the conversion from agent input to transmitted telepatheme; or in the conversion from received telepatheme to perceptual response. If it is the case that, as Price suggests, in telepathy the agent's mind is in direct 'contact' with the mind of the percipient, then of course there is no 'channel' between transmitter and receiver in the sense of ordinary communication theory. This is a highly paradoxical view which could only be accepted if it were really true that it was *impossible* to detect any influence passing between the brain of the agent and that of the percipient. *Prima facie*, as we have seen in the quotation from Eccles' *Scientific American* paper, the seat of the agent's consciousness is located in his brain, which is separated in space from that of the percipient; and therefore some channel or mode of spatial interconnection must be involved. If this is admitted then, as I have already said, in connection with Fisk's results quoted above, to be fair to the possibility of a physical explanation we must allow for possible 'ducting'.

THE FEASIBILITY OF A PHYSICAL THEORY OF ESP

In the case of the 'ducting' of high frequency radio waves these propagation conditions turn on the state of the Earth's atmosphere which acts as a reflector, so that the channel is more like a wave-guide than free space, and the inverse square law does not apply. Furthermore, in the case of physical communications, we have also to consider the remarkable techniques which exist for detecting and amplifying signals of very low power in the aerial, in such a way that the signals delivered to the loudspeaker—the analogue of consciousness—are delivered with sufficient strength to reproduce nearly all the original information content. It is perfectly possible that in the unconscious levels of the human brain there exist systems of selective detection and amplification by automatic gain control of weak incoming signals. Finally, the incoming signals might well be physical in the sense of being an influence or disturbance of the kind treated by physics without consisting of electro-magnetic waves. Indeed, what we know of the physical conditions operating in a living human brain, and the power it generates, make it exceedingly improbable that such a physical signal is electro-magnetic in nature. For the potentials detected by the electro-encephalograph at the surface of the brain, even during an epileptic fit, could only generate exceedingly weak signals of at most 10^{-18} watts, and are of very low frequency of less than 100 cycles a second. For all these reasons I find myself unable to accept Price's third assumption, which I have discussed at length because of its basic importance.

But it may still be said that, if the signal from the agent to the percipient in telepathy is to be any form of physical impulse or influence, it should be possible to intercept and detect it by a suitable instrument placed somewhere between agent and percipient. This brings us back to the first and second of Price's assumptions to which I referred on page 227 above. I shall now, therefore, turn to consider the question of what sort of detector mechanism might be involved, so that a physical influence which is incapable of being registered on one of the usual field strength meters, can yet be received and interpreted by the brain of the percipient. Here I shall quote from some remarks put forward by Eccles in his Waynflete Lectures at Oxford.[11] According to Eccles the

[11] Published as *The Neurophysiological Basis of Mind* (London: Oxford University Press, 1953).

cerebral cortex is the seat of dynamic spatio-temporal patterns of activity of immense complexity and delicacy. He says:

> There are some ten thousand million neurones in the human cerebral cortex and each is a node in the network whose strands are woven from the numerous processes (dendrites and axons) that provide the multiple synaptic contacts. Each node would be the convergent point of scores of paths and each in turn would project to scores of other nodes.[12]

These pathways connect neurones via little attachments called 'synaptic knobs', which are themselves minute in size, each being of the order of one micron (one millionth of a metre) in diameter and weighing about 5×10^{-13} grammes. These synaptic connections conduct to each neurone the discharges from other neurones in the network; and, as Eccles says, 'at each node the time of discharge is determined by the microscopic spatio-temporal pattern of the synaptic bombardment on its surface from millisecond to millisecond'.[13] (A millisecond is one thousandth of a second).

Eccles postulates nodes in this vast neuronal network, which are poised at 'critical' levels of excitability; and he claims that in this condition they can function as 'a "detector" that has a sensitivity of a different kind and order from that of any physical instrument' in relation to 'small influences'. In regard to the latter he makes the interesting suggestion:

> Suppose some small 'influence' were exerted at a node that would make a neurone discharge an impulse at a level of synaptic excitation that would otherwise have been just ineffective, that is, in general to raise the probability of its discharge.[14]

Eccles points out that the resultant effect of such a modification of probability of discharge, leading to the firing of a single 'unlikely' neurone, might be to induce the abnormal firing of hundreds of thousands of neurones (through a sort of chain reaction) within a very short period of about 20 milliseconds. The point is that this cascade of hundreds of thousands of firing neurones, would (a) be sufficient to produce an overt piece of behaviour; and (b) be solely due to the abnormal firing of a single 'critically poised' neurone in the first instance, which neurone would not

[12] *The Neurophysiological Basis of Mind*, p. 268.
[13] *ibid.*, p. 269.   [14] *ibid.*, pp. 273-4.

THE FEASIBILITY OF A PHYSICAL THEORY OF ESP

have fired at that particular time without the reception of the 'small influence'.[15]

Eccles based a theory of the neurophysical mechanism of the human will on these concepts, involving the notion of the human *brain* as a uniquely sensitive detector of 'small influences'. He says:

> Thus the neurophysiological hypothesis is that the 'will' modifies the spatio-temporal activity of the neuronal network by exerting spatio-temporal 'fields of influence' that become effective through this unique detector function of the active cerebral cortex. It will be noted that this hypothesis assumes that the 'will' or 'mind-influence' has itself a spatio-temporal patterned character in order to allow it this operative effectiveness.[16]

Though this theory was introduced to account for the initiation of *conative* processes of willing it can clearly be applied equally to *cognitive* processes of thinking and perceiving.

It is worth stressing the reason why Eccles suggests that the human brain's cortex can function as so sensitive a detector of influences, for if his argument is valid it can be used to show that another of Price's contentions is of no decisive significance against the physicalist attempts to give a physical explanation of ESP. I am referring to Price's remark that 'no explanation of that kind seems to be feasible. Such physical radiations, if they exist, ought to be detectable by physical instruments. It ought to be possible to intercept them en route. . . .' But it depends, of course, whether the kind of instruments Price has in mind would be sufficiently sensitive to detect physical signals of the kind which Eccles has in mind. Eccles explicitly says he is considering large networks of cortical neurones which are 'at a relatively high level of excitation' being 'subjected to an intensive synaptic bombardment'. As was said earlier, each of these neurones is a highly complex cell structure, which has about one hundred contacts with other neurones, via fibres which terminate at a peculiar sort of junction called a synapse. For a neural impulse to get through a synapse it is usually necessary for a number of other impulses from other neurones to arrive practically simultaneously. The synapse then integrates these impulses and, when their sum reaches a certain critical threshold, the neurone to which they are

[15] *The Neurophysiological Basis of Mind*, p. 276.   [16] *ibid.*, pp. 277-8.

connected fires off its own impulse. When the neurone is very near its 'flash over point' Eccles conceives it to be capable of discharging by a very small change occurring at just the right moment. Now this very small change could be due to a chance fluctuation, when the effect would be a spontaneous discharge of the kind which the communications engineer would call 'noise'—indeed, the concept of 'biological noise' is now a well-accepted idea in neurophysiology, where it is generally attributed to random variations in the ionic concentrations of cell metabolites.[17] On the other hand, it could also be due to an external influence acting on one of the knobs of the synapse, which is as we have seen only about one micron in diameter. Then such a disturbance can carry information about an external event, since it can give rise to an adaptive response in the neurone chain that is in fact appropriate to the external situation which has given rise to the initial disturbance. In these circumstances the firing of the critically poised neurone chain is not attributable to 'noise' but is an instance of genuine detection of a signal in the sense of information theory in physics. This is evidently the kind of situation which Eccles has in mind when he talks of the cerebral cortex as being a sensitive detector of small influences, but he gives us no insight into the nature of this influence. On the other hand, he definitely commits himself to saying that the influence must itself have a spatio-temporal pattern; and that it must be located in the *physical* space and time in which the cortex upon which it operates is located. Most people would, I think, hold that anything located in physical space must be itself physical, and be subject ultimately to the laws of physics, even if it is not *directly* observable by ordinary physical instruments. There are, indeed, plenty of instances in modern physics of entities which are postulated to be subject to its laws and which are not directly observable or measurable: such as, for example, the so-called 'virtual' particles of quantum electrostatics, according to which all charged particles such as electrons are surrounded by unobservable virtual photons.

I shall have more to say shortly about the significance of virtual particles for the physics of ESP. The point I wish to make now

[17] See, for example, the paper entitled 'Fluctuations in Neural Thresholds' by Lawrence S. Friskopf and Walter A. Rosenblith in *Information Theory in Biology*, ed. by H.P. Yockey *et al.* (Oxford: Pergamon Press, 1958), espec. p. 160.

THE FEASIBILITY OF A PHYSICAL THEORY OF ESP

is that all instruments used for recording, measuring, and observing today give readings *directly* in terms of changes of state involving *mathematically real* physical energy. Such mathematically real physical energy can, of course, be of many different specific kinds such as electro-magnetic, mechanical, chemical, etc. But these varieties are all specific forms of what is technically known as mathematically 'real'—as opposed to mathematically 'imaginary' —energy. What I want to suggest is that the kind of influences which might serve to trigger off a neuronal chain-reaction, starting in a critically poised neurone as conceived by Eccles, could be changes involving mathematically 'imaginary' energy. This suggestion needs elucidation because the distinction between mathematically 'real' and mathematically 'imaginary' quantities may not be familiar to everyone.

The simplest way of approaching the idea of a mathematically 'imaginary' quantity is to consider the case of an ordinary equation, such as $x^2 = -1$. We all remember that the square of any ordinary or 'natural' number, such as any of the integers 1, 2, 3, etc., must invariably be a positive number. The reason is, of course, that if you square a negative natural number, i.e. multiply it by itself, the product of the two 'minus' signs yields a 'plus' sign, and thus the result is a positive number. So what kind of number can it be which, when squared or multiplied by itself, produces a negative sign as in the equation $x^2 = -1$? It was in this sort of way that mathematicians first came up against the problem of the so-called 'imaginary numbers', which they invented for the express purpose of providing roots to quadratic equations in terms of numbers whose squares were negative. It was not till the nineteenth century that these numbers received adequately rigorous treatment, when it was shown by Gauss and others how they might be incorporated in an extended number system, involving ordered couples of 'real' numbers, one of which played the part of the pure 'imaginary' number. We need not concern ourselves with the formal mathematical properties of such numbers, beyond noting that their introduction into expressions relating to physical quantities, such as energy or mass, or time, involves adding an additional dimension into the manifold in which such quantities are conceived to be located. For the pure 'imaginary' number $\sqrt{-1}$ is represented in the complex number system by the number (0, 1) which is composed of the ordered pair 0 and

1 of real numbers. Similarly, the 'real' number 1 is represented by the complex number (1, 0) and the number $1+\sqrt{-1}$ by the number (1, 1). So if we say that a particle has an imaginary energy of $\sqrt{-1}$ ergs we would write this (0, 1) ergs and have to represent it graphically on a two-dimensional diagram as follows:

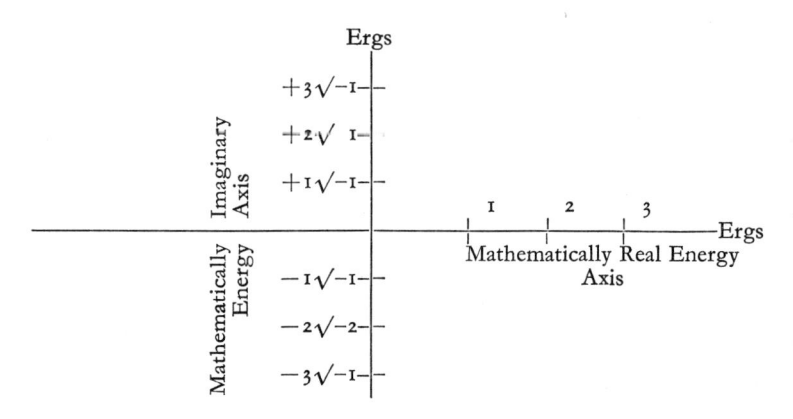

```
                        Ergs
                    +3√−1 ─┤
                    +2√−1 ─┤
      Imaginary
         Axis       +1√−1 ─┤
                           │      1    2    3
                    ───────┼──┼────┼────┼────── Ergs
                           │   Mathematically Real Energy
  Mathematically    −1√−1 ─┤              Axis
       Energy
                    −2√−2 ─┤
                    −3√−1 ─┤
```

Figure 3: Mathematically Imaginary

Now it will be obvious that if one is only using instruments designed to be influenced by energy and momentum changes, which are expressable in terms of mathematically 'real' numbers, one will not be able to detect directly any influence which takes effect through an imaginary energy or momentum change. We cannot expect to detect directly any change of energy which is only expressible in terms of a mathematically 'imaginary' number, a number of the form ($\sqrt{-1}x$) ergs where x is a mathematically real number. But it does not follow that such influences do not occur and have a real physical effect. Indeed, quantum physics explicitly recognizes this and admits that such changes must be supposed to occur with great frequency. A Russian physicist, Ya. P. Terleiski, has pointed this out very clearly in the following passages taken from one of the leading journals of physics in the U.S.S.R.

> 4. Consequently, particles moving with a velocity greater than the velocity of light or, in other words, particles with imaginary intrinsic mass are not forbidden categorically as if they were *physically* unreal. Forbidden only is the process in which the emission of such

particles is systematically repeated and associated with an increase of entropy of the radiator. The emission or absorption of such particles may be either a process of a purely fluctuational character, or may take place reversibly and, in general, without a change in the entropy of the radiator or receiver.... Consequently, a particle with a real mass can radiate and absorb particles of imaginary mass, with only the kinetic energy of the ensuing motion being changed without a change in the intrinsic mass.[18]

Moreover, according to current quantum theory, whenever a particle or other physical entity is interacting with an electromagnetic or nuclear field of force the interaction takes place through the medium and intervention of the so-called 'virtual' particles, which can have mathematically imaginary mass or momentum. As a well-known physicist Professor W. H. Watson has said, 'In the interactions between fields we consider imaginary values possible for the momenta of the particles created.'[19] Thus, according to current orthodox quantum physics every material particle of mathematically real mass can be surrounded by a gas of particles of imaginary mass, the so-called 'virtual' particles, carrying mathematically imaginary energy and momenta; and these 'virtual' factors can play an essential part in the ordinary physical processes.

Virtual particles with imaginary energies and rest-masses indeed contribute some of the characteristic features of the behaviour of ordinary particles such as electrons and positive ions, which have mathematically real energies and rest-masses But no one would suggest that these virtual particles (or any entity with mathematically imaginary physical characteristics such as imaginary mass or momentum) could be *directly* detected in any ordinary piece of physical apparatus. Their occurrence has to be *inferred* from the experimental results interpreted in the light of the fundamental physical principles. But this fact about virtual particles does not make them any less *physically* real or *material* than an electron or a neutrino. They are still an essential feature in the physical world as envisaged by the physicist. Modern physics is, in fact, well accustomed to using quantities and operations which represent components of the physical system under

[18] *Doklady Akademii Nauk SSR*, Vol. 133, No. 2, pp. 329–32.
[19] *Understanding Physics Today* (London: Cambridge University Press, 1963), p. 151.

observation, but which are either undetectable with present apparatus or intrinsically unobservable according to the tenets of physics itself. Of the latter type of entity I have already mentioned the virtual photons, but an equally good example is the Dirac 'infinite sea' of negative energy electrons in his theory of the positive electron.

A striking example of the former sort of entity is the neutrino. The *concept* of this particle (which has no positive physical characteristic except intrinsic spin) was first seriously proposed by Fermi in 1932, following up an *ad hoc* suggestion of Pauli, who originally introduced the notion solely to preserve the principle of the Conservation of Angular Momentum in elementary particle reactions. But Fermi's first paper on the subject which suggested how the neutrino might be used to explain the facts of beta decay, was rejected by the scientific journal *Nature*, on the grounds that a hypothesis involving an apparently undetectable entity was of no interest to its readers. It took approximately thirty years, and huge strides in technology, to make possible the 'observation' of the neutrino; and even now it is only detectable by very indirect methods. Yet every competent physicist today is convinced of the particle's existence and of the essential part it plays in the so-called 'weak' interactions in experimental physics. According to Price's criteria, however, it was once a non-starter as a genuine physical entity.

I have brought in this discussion of the neutrino and the virtual photons as instances showing that orthodox physics provides for the occurrence of entities which are not directly observable. These entities (which include the particles of imaginary mass and energy mentioned by Terletski as well as the negative energy states of electrons of real mass mentioned by Dirac) are essential parts of the modern physicist's mathematical formulation of the physical world.

Thus modern theoretical physics makes abundant provision for the kind of influence which is not directly observable by instruments measuring changes in mathematically real positive energy and momentum. This point is sufficient to refute Price's contention, that if telepathy is due to a physical influence such influence must be directly observable in physical instruments. It also serves to provide a definite basis for the hint thrown out by Eccles, of the possibility of a 'small influence' which can serve to operate

THE FEASIBILITY OF A PHYSICAL THEORY OF ESP

upon the critically poised neurones of the human brain, so as to trigger off an observable effect in the central nervous system. We can go farther, and suggest an *indirectly* observable effect of such interaction between these neurones and particles of imaginary mass and energy. For one consequence of such interaction should be a directly observable phase shift in the characteristic waves detectable in the electro-encephelograph.

Those interested will find a fuller account of the process in my paper 'Time and Extra-Sensory Perception',[20] especially in Section III entitled *Two-dimensional Time and Probability Patterns*. I show there that it is possible to give a physicalist account of certain paranormal psychological phenomena, by the introduction of a two-dimensional theory of time. I also show how such a theory can give certain predictions capable of being tested by the kind of experimental techniques developed by the brain physiologists. I will only say here that the introduction of a two-dimensional time is a natural corollary of the admission of imaginary values for energy in physics. The reason is, briefly, that according to quantum theory the quantities 'energy' and 'time' in physics have a particularly close connection with each other, in accordance with Heisenberg's Uncertainty Principle.

There is an interesting development of this conception of Heisenberg's Uncertainty Principle in the work of the Russian physicists V. A. Folk and N. S. Krylov, who showed on this basis how to derive the fundamental law of the time decay of an excited state of a physical system in quantum mechanics.[21] I have shown elsewhere how this two-dimensional time theory—the theory of time as a complex rather than a real variable in the mathematician's sense—can be adapted from quantum physics and applied to elucidate the phenomenon of ostensible precognition.[22] I shall not be able to go into the details of this theory here; but for my present purpose it will be sufficient to summarize very briefly some of the conclusions which are relevant to a physical explanation of ESP., and which have implications for a new but practical programme of parapsychological experiment.

According to this theory, time in physics and psychology has both a real component (in the mathematical sense) in which

[20] Published in the *Proc. S.P.R.*, Vol. 54, Pt. 196, 1965.
[21] 'Soviet Physics', *JETP*, Vol. 17, No. 93, 1947.
[22] *Proc. S.P.R.*, Vol. 54, Pt. 196, 1965.

actual happenings or events are dated; and an imaginary component (again in the mathematical sense) in which the objective probabilities of different possible outcomes of events are ordered. Certain of these possible outcomes or eventualities (those which have an outstandingly high probability of happening determined by the physical conditions) are conceived as being able to give rise to a kind of datum I call a 'Precast' of an event. Such data are what a person is directly and non-inferentially aware of when ostensible precognition occurs. These *objectively* highly probable precasts of events, which are very *likely* but are not *certain* to happen in the future, may have relatively low *subjective* probabilities for the conscious mind of the person who has the experience (or behaves as if he had the experience) of ostensible precognition.

The objectively probable precasts of events are conceived as spread out in a quasi-temporal order around actual events, this order being represented by the imaginary time variable, at each moment of the mathematically real time variable. Besides involving a two-dimensional time, the theory also involves the quantum-mechanical concept of objective probabilities as *propensities*, which influence the outcomes of eventualities in a statistical way as conceived by Karl Popper. Such objective (physical) probabilities 'incline without necessitating' (Leibnitz) the occurrence of individual events, many of which do not involve human beings at all; for example, the ejection of an alpha particle from a radioactive nucleus.

Epistemologically, ostensible precognition is to be conceived as an awareness by a human being of objective probabilities which serve as precasts of future event possibilities. Such human responses to objective probabilities, initially at the subconscious level, can manifest themselves behaviourally (as in the Soal–Shackleton experiments); or they can trigger off introspectibly conscious states, such as visual or auditory images, or sensory hallucinations found in the 'spontaneous' cases. Analogous responses to the immediate probabilities, as directly apprehended, are to be found in many ordinary perceptual situations requiring rapid bodily adjustment of an anticipatory kind (as in playing fast ball games such as tennis or racquets).

A distinguished physicist and fellow-contributor to this book, Professor Margenau of the Sloane Laboratory Yale University, himself responsible for major contributions to the quantum theory

THE FEASIBILITY OF A PHYSICAL THEORY OF ESP

of the Van Der Waals forces, has made certain observations about multi-dimensional time theories in physics, which call for some comment. Margenau says:

> An artifact occasionally invoked to explain precognition is to make time multi-dimensional. This allows a genuine backward passage of time, which might permit positive intervals in one time direction to become negative ('effect before cause') in another. In principle this represents a valid scheme, and I know of no criticism that will rule it out as a scientific procedure. If it is to be acceptable, however, a completely new metric of space-time needs to be developed, a metric which will account not only for the facts to be explained, but also for the known laws of physics.

This is a perfectly valid point, but in fact the answer to it is already to hand. It is to be found first in Eddington's pentadic theory of space-time, which received its clearest expression in his *Fundamental Theory* which was published after his death by Sir Edmund Whittaker.[23] Eddington's theory was based upon a space-time manifold in which two dimensions are time-dimensions and three are space-dimensions. He was able to show how all the usual results of relativity and quantum theory could be fitted into such a manifold though, owing to a certain metaphysical confusion about time, he failed to accord his second time dimension the full physical significance it deserved. But, more recently, an even clearer proof of the feasibility of deriving all the usual results of physics, from a pentadic theory of space-time identical with Eddington's is to be found in a book by Professor H. T. Flint, formerly Professor of Physics at London University. This book is entitled *The Quantum Equation and The Theory of Fields*[24]. Flint develops a five-dimensional theory in which all the usual results of relativity and quantum theory are deduced; and the identity of its pentadic (two-dimensional time) structure with Eddington's is manifest—see especially Chapter 7, and equation 7.23, on p. 140, in particular.

Apart from the particular case of ostensible precognition the quantum mechanical theory of complex time and energy provides a possible physical causal mechanism for *general* ESP. For this theory allows *physically real* processes involving particles of

[23] London: Cambridge University Press, 1946.
[24] London: Methuen, 1966.

mathematically imaginary mass to occur reversibly, and without frictional loss of energy across space (no frictional or other *irreversible* dissipation of energy can occur in interactions between particles of imaginary mass or energy as has been pointed out by the Russian physicist Y. P. Terletski). The physical basis for ESP would then consist in the interactions between the ordinary particles of mathematically real rest mass (such as the molecular micro-constituents) of a human brain and a gas of particles of mathematically imaginary mass, such interactions triggering the unexpected firing of Eccles' 'critically poised' neurones. But for Eccles' vague and undefined notion of a 'small influence', whose complete lack of specification has been rightly criticized as a defect in an otherwise admirable scientific statement, I wish to substitute the precise physical concept of 'perturbation due to interaction between particles of real and imaginary rest mass'.

This approach at once suggests an entirely new experimental programme of ESP research with the particular object of elucidating its physical basis. The object would be to test the hypothesis that general ESP and telepathy in particular is due to the stimulation of ordinary particles in the brain ('critically poised' neurones), by patterns of particles of imaginary rest mass, which should lead to correlated shifts in the phases of the relevant E.E.G. waves of two persons en rapport.

The first step should be enlarged experiments along the lines conducted by the United States Air Force, using rapid and efficient randomization and recording procedures, designed to pick out significantly above chance scorers in guessing experiments. Such subjects should then be put through E.E.G. tests designed to reveal the occurrence of phase-shifts predicted by the theory.[25] Such experiments should be designed so that even if significant indications of ESP are totally lacking useful data can be obtained for the ordinary experimental psychologist working in the field of Reaction-Choice or Decision Times, or in the field of subthreshold stimulation. These fields are all of direct practical importance both to commercial and government undertakings. That such by-products can be obtained from well-designed experiments can be seen from the reports of the United States Air

[25] For further details as to the hypothesis, its predictions and specifically how to test them, see the paper on ESP and Time in *Proc. S.P.R.*, Vol. 54, Pt. 196, by the author.

Force experiments which, in fact, produced data on choice reaction and decision-making times.

Data on the effects of stimuli above the physiological threshold, but below the conscious awareness threshold, are of great interest both to practical psychologists and to the advertising industry. It is not difficult to design experiments which will test capacities for both 'normal' sensory reception of sub-threshold stimuli and for 'paranormal' extra-sensory perception. A detailed description of just such an experimental set-up is to be found in the work of an American experimental psychologist reported in his 'Discrimination without Awareness'.[26] There is, in fact, an immense field for useful collaboration between the experimental psychologist, physiologist, and parapsychologist which has been made possible by modern apparatus. To get this collaboration started, however, one needs to test a specific hypothesis with predictions involving phenomena from these different fields. It is in this spirit that I offer the foregoing sketch of a physicalist theory of ESP which any interested reader will find elaborated in much greater detail in the paper published in the *Proceedings S.P.R.* to which I have referred.

In closing I would like to emphasize one rather curious consequence which would follow if the hypothesis that ESP is attributable to interactions involving particles of mathematically imaginary mass is accepted. It is implied in the remarks I have quoted from the Russian physicist Y. P. Terletski:

> Particles moving with imaginary intrinsic mass are not forbidden categorically as if they were physically unreal. Forbidden only is the process in which the emission of such particles is systematically repeated and associated with an increase in entropy of the radiator. The emission or absorption of such particles may be either a process of a purely fluctuational character or may take place reversibly and in general without a change in entropy of the radiator or receiver.

The significance of these remarks for the theory of ESP sketched in this chapter is that the theory rules out the possibility of harnessing ESP as a new means of systematic communication analogous to ordinary radio. The most one could expect to do would be to develop a system for recognizing the circumstances

[26] *Amer. J. Psychol.*, Vol. 52, pp. 562–78.

surrounding occurrence of the disturbances of a 'fluctuational character' and exploiting the fluctations when they occur spontaneously. For it is the essence of a 'fluctuation' that it is from the standpoint of physics a chance phenomenon for which one cannot give a fixed rule prescribing its occurrence in advance. Further, as Terletski makes clear, the only *regular* and *predictable* absorption and emission of particles of imaginary mass are such that they can occasion no change of entropy in either the radiator or the receiver; and such transactions without change of entropy cannot carry information according to the accepted principles of physical communication theory. This implication of the theory may be the explanation of the curiously elusive, unpredictable, and apparently unrepeatable nature of ESP phenomena, which manifest themselves statistically in controlled experiments. It would follow that such acts of ostensible communication which occur in the well-attested instances of telepathy and/or precognition in the spontaneous cases were of the nature of unpredictable 'fluctuations', representing a strongly felt need in the subject that conveyed information by the exchange irreversibly of particles of imaginary mass between agent (radiator) and percipient (receiver). They would not then be attributable to *voluntary decisions* on the part of the agent; unless we are prepared to say that what appears to be a 'voluntary decision' on a particular occasion is, in fact, an undetermined fluctuation of the material particles of the brain, which would not (even in principle) be predictable in advance by physics. Such a fluctuation would, of course, be in principle completely describable in the language of physics *after it had occurred*, though it would not be determinate or predictable in advance.

XI

PSYCHOANALYSIS AND PARAPSYCHOLOGY

Emilio Servadio

The psychoanalytical approach to the problem of parapsychology was held up and hindered, firstly, by a lack of strict methodology in many past researches and works dealing with 'metapsychic' or 'paranormal' phenomena (a subject open for many years to a whole crowd of amateurs and enthusiasts, 'spiritualists', and so on), and secondly, by the fact that nearly all psychoanalysts and the great majority of scientists feared a 'return to magical thinking'. The conclusion that parapsychological investigations can be carried out in strict compliance with the most exacting scientific tenets and have nothing to do either with superstition or occultism was only reached fairly recently.

Freud's work

Freud's contributions to the studies in question are dated from 1899 to 1933.[1] His first article on telepathy was published in

[1] S. Freud, 'A Premonitory Dream Fulfilled' (1899), in *Standard Edition of the Complete Psychological Works*, Vol. V; idem, 'The Psychopathology of Everyday Life' (1904), Ch. XII, in *S.E.*, Vol. VI; idem, 'Dreams and Telepathy' (1922), *S.E.*, Vol. XVIII; idem, 'The Occult Significance of Dreams' (1925), in *S.E.* Vol. XIX; idem, 'New Introductory Lectures on Psycho-Analysis' (1933), Ch. 2, Lecture XXX, in S.E., Vol. XXII. *Note*: The six works quoted above, as well as the main analytical contributions (up to 1952) to parapsychological problems in so far as they were related to analytic work, are included in the following anthology: G. Devereux (ed.), *Psychoanalysis and the Occult* (New York: International Universities Press, 1953).

1922.[2] In it Freud developed the extremely fruitful idea that telepathic perceptions, if they occur, could easily undergo distortions typical of the workings of a mind dominated by unconscious desires (displacement, condensation, symbolization, 'censorship', and defence mechanisms in general). A man had written to him that he had dreamt his second wife had given birth to twins. The following day he heard that his daughter by his first wife had had twins, the same night, although she had only expected to be delivered about a month later. According to Freud, if one admits that the dream was telepathic, the divergence between its manifest content and what really happened could very well be explained, from a psychoanalytical point of view, by the substitution of the dreamer's wife for his daughter. He might have repressed the desire to be the father and not the grandfather of the child that was about to be born. The telepathic perception of the delivery might have endangered the repression of this incestuous desire and the censorship had consequently transposed and masked it by representing it as the desire to have a child not from his daughter but from his wife.

In this way, Freud concludes, psychoanalysis would be able to recognize the telepathic contents included in other dreams and that had undergone a similar censorship and deformation. Moreover, the same principle might be applied to telepathic perceptions when awake, and to their apparent contents. Freud seems to admit, however, that there may also be 'pure' (that is to say undistorted) manifestations of a paranormal knowledge of events. If this is so, he writes, it would be wrong to speak of telepathic 'dreams' in this case as they would actually be 'telepathic events occurring during sleep'.

In a short article that appeared in 1925,[3] Freud took up these considerations again, adding that, in his opinion, there were probably unconscious emotional factors common to the persons involved in the telepathic communication that analysis would be able to reveal.

Lastly, in Chapter II (Lecture 30) of his *New Introductory Lectures on Psychoanalysis*,[4] Freud deals directly with the question of telepathic phenomena during analysis, describing in great detail

[2] 'Dreams and Telepathy', *loc. cit.*, above.
[3] 'The Occult Significance of Dreams', *loc. cit.*, above.
[4] *loc. cit.*, above.

PSYCHOANALYSIS AND PARAPSYCHOLOGY

an example from his own case-book: the 'case of Dr. Forsyth'. Freud tells us how his patient P had perceived his doctor's transient indifference and also—but this by the possible channel of telepathy—Freud's preference for a new patient from England. In the last lines of this work, Freud stresses the archaic and regressive character of telepathic phenomena. Telepathy, he writes, 'may be the original archaic means through which individuals understood one another; a means pushed into the background, in the course of phylogenetic development, by a better method of communication, that is to say by signs perceived by the sensory organs. But these older methods of communication may have survived in the background and may still manifest themselves under certain conditions'.

Other contributions prior to 1939

Following in Freud's footsteps, several psychoanalysts—Hollós [5] and Servadio [6] among the first—endeavoured to establish first of all the psychodynamic conditions (such as can be observed especially during analytical treatment) under which certain parapsychological phenomena (telepathy and perhaps also precognition) seem to be produced. Their example was followed by Fodor [7], Ehrenwald, [8] and some others. The following points had been fairly well ascertained before 1939 and were taken up again and developed after the 1939-45 war. Firstly, as Freud had pointed out, cases of telepathy during analysis revealed the full importance of emotional factors in the composition of a telepathic incident. These factors do not only derive from conscious or preconscious emotions such as enthusiasm, fear, ambition or rivalry, but are connected with deep primitive impulses and conflicts: scopophilia, exhibitionism, weaning- and castration-complexes, incestuous desires, sadomasochistic and scatological fantasies, and so on. Hollós and Servadio reported their patients' telepathic dreams or fantasies. This material showed that there was something in common—or rather something complementary—between

[5] I. Hollós, 'Psychopathologie alltäglicher telepatischer Erscheinungen' in *Imago*, Vol. XIX, 1933.
[6] E. Servadio, 'Psychoanalyse und Telepathie', in *Imago*, Vol. XXI, 1935.
[7] N. Fodor, 'Telepathic Dreams', in *Imago*, Vol. III, 1942.
[8] J. Ehrenwald, 'Telepathy in the Psychoanalytic Situation', *Brit. J. Med. Psychol.*, Vol. XX, 1944.

the patient's problems and those troubling the analyst at the time, and that the patient 'unmasked' the analyst, so to speak, by means of paranormal perception. Ehrenwald[9] stressed the fact that the phenomena in question occurred when there was a defect or a weakness in the protection and defences of the psychic or psychophysiological Ego that allowed the transitory restoration of more primitive mechanisms or means of communication. All of them confirmed Freud's views concerning the deformations the parapsychic contents can undergo before reaching the subject's consciousness.

Contributions after 1939—present concepts

The most remarkable contributions to the problems of psychoanalytical parapsychology were made after 1939 by Fodor,[10] Ehrenwald,[11] Eisenbud,[12] and Servadio,[13] to whose names must

[9] J. Ehrenwald, *Telepathy and Medical Psychology* (New York: Norton, 1948); *idem*, 'Presumptively Telepathic Incidents during Analysis', *Psych. Quart.*, Vol. XXIV, 1950.

[10] N. Fodor, 'Telepathic Dreams', *American Imago*, Vol. III, 1942; *idem*, 'The Psychoanalytic Approach to the Problems of Occultism' in *J. Clin. Psychopathol.*, Vol. VII, 1945; *idem*, 'Telepathy in Analysis', *Psych. Quart.*, Vol. XXI, 1947; *idem*, *On the Trail of the Poltergeist* (New York: The Citadel Press, 1958); *idem*, *The Haunted Mind* (New York: Helix Press, 1959).

[11] J. Ehrenwald, 'Telepathy in the Psychoanalytic Situation', *Brit. J. Med. Psychol.*, Vol. XX, 1944; *idem*, *Telepathy and Medical Psychology* (New York: Norton, 1948); *idem*, 'Presumptively Telepathic Incidents during Analysis', *Psych. Quart.*, Vol. XXIV, 1950; *idem*, *New Dimensions of Deep Analysis* (London: Allen & Unwin, 1954); *idem*, 'Induction télépathique et complaisance doctrinale', *La Tour Sainte-Jacques*, No. 6–7, 1956; *idem*, ' "Non Euclidian" Models of Personality', *Int. J. Parapsychol.*, Vol. I, 1959.

[12] J. Eisenbud, 'Telepathy and Problems of Psychoanalysis', *Psychoanal. Quart.*, Vol. XVI, 1946; *idem*, 'The Dreams of Two Patients in Analysis interpreted as a Telepathic "Rêve à deux" ', *Psychoanal. Quart.*, Vol. XVII, 1947; *idem*, 'Analysis of a Presumptively Telepathic Dream', in *Psych. Quart.*, Vol. XXII, 1948; *idem*, 'On the Use of the Psi Hypothesis in Psychoanalysis', *Int. J. of Psycho-Anal.*, Vol. XXXVI, 1955; *idem*, 'Time and the Oedipus', *Psychoanal. Quart.*, Vol. XXX, 1956.

[13] E. Servadio, 'Psychoanalyse und Telepathie', *Imago*, Vol. XXI, 1935; *idem*, 'La baguette des sourciers: essai d'interprétation psychanalytique', *Revue Française de Psychanalyse*, Vol. VIII, 1935; *idem*, 'Psychoanalysis and Yoga', *Bulletin of the Bombay Medical Union*, No. 3/4, 1940; *idem*, 'Psychologie des profondeurs et parapsychologie', *La science et le paranormal* (Paris: Institut Metapsychique International 1954); *idem*, 'A Presumptively Tele-

be added those of Balint,[14] Gillespie,[15] Rubin,[16] Pederson-Krag, *et al.*,[17] who also wrote on the subject.

Fodor had been interested in 'telepathic dreams' since 1942 and held that certain dreams cannot be understood unless one accepts the phenomenon of telepathy. Of more recent date[18] is the complete publication of his psychoanalytical study (made before the last war) of a Poltergeist case. The case, that is to say, of a woman who found herself the butt of more or less violent 'paraphysical' phenomena of an alarming nature. Freud had read Fodor's work and approved of the spirit and method of it.

Two books and several essays by Ehrenwald[19] have illustrated the 'parapsychological dimension' in interpersonal relations, particularly during analysis. Ehrenwald has also advanced some brilliant hypotheses concerning the possibility of (telepathic) 'heteropsychic invasions' in certain schizophrenics (especially of the paranoid type) and of the pathogenic function of some extrasensory relations between mother and child in the first months of

---

[14] M. Balint, 'Notes on Parapsychology and Parapsychological Healing', *Int. J. Psycho-Anal.*, Vol. XXXVI, 1955.

[15] W. H. Gillespie, 'Extrasensory Elements in Dream Interpretation' in *Psychoanalysis and the Occult*, 1953, ed. by G. Devereaux (New York: International Universities Press, 1953).

[16] S. Rubin, 'A Possible Telepathic Experience during Analysis', in *Psychoanalysis and the Occult*, cited note 15 above.

[17] G. Pederson-Krag, 'Telepathy and Repression', *Psycho-Anal. Quart.*, Vol. XVI, 1947.

[18] N. Fodor, *On the Trail of the Poltergeist* (New York: The Citadel Press 1958).

[19] See note 11 above.

---

pathic-Precognitive Dream during Analysis', *Int. J. Psycho-Anal.*, Vol. XXXVI, 1955; *idem*, 'Le conditionnement transférentiel et contre-transférentiel des événements "psi" au cours de l'analyse', *Acta Psychotherapeutica, Psychosomatica et Orthopaedogogica*, Supplement to Vol. III, 1955; *idem*, 'Telepathy: a Psychoanalytic View', *Tomorrow*, Vol. IV, 1956; *idem*, 'Freud et la parapsychologie', *Revue Française de Psychanalyse*, Vol. XX, 1956; *idem*, 'État présent de la parapsychologie psychanalytique', *La Tour Saint-Jacques*, No. 6/7, 1956; *idem*, 'Transference and Thought-Transference', *Int. J. Psycho-Anal.*, Vol. XXVII, 1956; *idem*, 'Magic and the Castration-Complex', *Int. J. Psycho-Anal.*, Vol. XXXIX, 1958; *idem*, 'Parapsychologie und Ungläubigkeitsreaktion', *Zeitschrift fur Parapsychologie*, Vol. II, 1958; *idem*, 'Telepathy and Psychoanalysis', *J. Amer. Soc. Psych. Res.*, Vol. LII, 1958; *idem*, 'The Normal and the Paranormal Dream', *Int. J. Parapsychol.*, Vol. IV, 1962.

its life. Recently Ehrenwald[20] has suggested an 'open' (or 'non-Euclidian') model of the human personality which, unlike the traditional 'closed' model, takes the parapsychological factor into account.

Eisenbud[21] has revealed the wide range of telepathic interactions not only between analyst and patient but also between two or more persons treated by the same analyst. Eisenbud, too, thinks that telepathy and precognition belong quite naturally to the realm of normal psychic functions and that the 'psi factors' may derive directly from the unconscious, just like any other kind of stimulus.

Servadio[22] has devoted his attention to describing the conditions that appear to favour telepathic phenomena, especially during analysis. In his opinion, these conditions are related to a particular situation of transference and counter-transference, in which the patient feels frustrated by a lack of attention on the part of his analyst, who is engrossed in problems of his own, whose direction and content closely resembled the problems worrying the patient. Other forms of expression and communication being blocked, the frustrated patient has recourse to a more direct and primitive means, such as telepathy or even precognition (in a dream or otherwise). He is thus able to penetrate the analyst's 'secrets', to show him he is acquainted with them, to divert the analyst's libido to his own advantage, to reproach him for his lack of love or his aggressiveness. The analyst, in turn, appears to find himself in the conditions most favourable to such manifestations when he is unconsciously 'withdrawing' too much from his patient's 'infantile' desires, and when he has worries and interests which are complementary to his patient's, but which divert his attention, or even when there is an element of hostile counter-transference.

Servadio has endeavoured to extend to what are referred to as 'spontaneous' cases of telepathy the psychodynamic principles he believes to have established with regard to 'psi' manifestations in analysis—starting from Freud's idea that transference and counter-transference phenomena are universal and that the analytical situation merely brings them to light and enlarges them. He has

[20] J. Ehrenwald, ' "Non Euclidian" Models of Personality', *Int. J. Parapsychol.*, Vol. I, 1959.
[21] See note 12 above.   [22] See note 13 above.

PSYCHOANALYSIS AND PARAPSYCHOLOGY

also tried to show that parapsychological phenomena are easier to understand if one conceives the conscious, individualized, and rational aspect of the personality as a recent conquest which is inseparable from a deeper and more archaic aspect in which there is neither rationality, nor conscience, nor individuality.

Balint,[23] who also stresses the transference and counter-transference premises of 'psi' events in analysis, has called attention to certain 'defences' of the analyst other than the mere repression or scotomization of phenomena. These defences are projection (the phenomena are produced by the patient and have nothing to do with the analyst) and idealization (the phenomena form an interesting scientific problem, but the underlying conflicts are ignored).

In their approach to parapsychological questions, psychoanalysts are clearly concerned above all with telepathic phenomena (particularly during analysis) and the problem of precognition, so that to a considerable extent they neglect the others. However we have mentioned Fodor's interest in Poltergeist cases[24] and may mention also Servadio's studies on dowsing[25] and yoga.[26] Some psychoanalysts have shown an interest in certain psychodynamic implications or 'quantitative' parapsychological researches (Rhine's school) and in a more thorough study of the personality of mediums, 'sensitives', healers, etc., as well as in the psychology of inter-human relations between the latter and their clients.

The psychoanalytical approach has undoubtedly helped parapsychology to make remarkable progress and this progress is being continued and consolidated. On the other hand, the acceptance of telepathic or extra-sensory communications considerably enriches the psychoanalytical views of the human personality, and of inter-individual relations between children and adults, as well as of the analytical situation itself and of psychoanalytical technique as such. The psychoanalytical study of parapsychological phenomena has, above all, made it possible to recognize that there is nothing supernatural or super-normal about these phenomena and to insert them gradually into a broader and more satisfactory view of mental life.

[23] See note 14 above.   [24] See note 18 above.
[25] E. Servadio, 'La baguette des sourciers: essai d'interprétation psychanalytique', *Revue Française de Psychanalyse*, Vol. VIII, 1935.
[26] *idem*, 'Psychoanalysis and Yoga', *Bulletin of the Bombay Medical Union*, No. 3/4, 1940.

XII

C. G. JUNG AND PARAPSYCHOLOGY

Aniela Jaffé

To Carl Gustav Jung parapsychology was not merely the subject for scientific research, theory, and experiment. His life was rich in personal experiences with spontaneous, acausal, or—to use the common term—paranormal phenomena. He seemed to be endowed with an unusual 'permeability' to events in the 'background' of the psyche. But that alone does not explain the extent of his experiences; his sensitivity to manifestations of the unconscious was supplemented by constant observation and study of nature, of objects, and of people. Given his close attention to the worlds of the psyche and of external reality, it is not surprising that he observed meaningful connections between the two, connections that would have been overlooked by a less sharp observer. Prophetic dreams and precognitions were no rarity in Jung's life. But whenever they occurred he noted them with surprise—one is tempted to say, with the awe due to the miraculous.

Jung's interest in parapsychology as a science began while he was studying medicine, that is, in the last years of the nineteenth century, when terms like 'somnambulism' or 'spiritism' were much in vogue. One of his old school friends, Albert Oeri[1], describes Jung's early interest in an essay which he dedicated to him on the occasion of his sixtieth birthday:

> I cannot deny that Jung underwent a severe test of personal courage when he studied spiritistic literature, did a good deal of experimentation in that field and stood by his own convictions unless they were modified by more careful psychological studies. He was up in arms when official science of the time simply denied

[1] Later Nationalrat (member of the Swiss Parliament) and editor of the daily newspaper *Basler Nachrichten*.

ANIELA JAFFÉ

the existence of paranormal phenomena, instead of investigating and explaining them. Thus spiritists like Zöllner and Crookes, whose doctrines he could discuss for hours, became for him heroic martyrs of science. Among friends and relatives he found participants for spiritistic seances . . . I enjoyed enormously listening to Jung's 'lectures' on this subject when I came to see him in his lodgings. His charming dachshund would look up at us so gravely, as if he understood everything, and Jung used to tell me that the sensitive animal sometimes whimpered pitously when some occult force manifested itself in the house.[2]

As Oeri indicates, Jung did not confine himself to reading literature on the paranormal. He began his own experiments, and during the years 1899 and 1900 organized regular seances. The medium was one of his cousins, a fifteen-year-old schoolgirl. At the beginning of this enterprise, which he later analysed in his dissertation 'On the Psychology and Pathology of so-called Occult Phenomena'[3] two such phenomena took place in the home he was sharing with his widowed mother and sister. A heavy walnut table, an old heirloom, split with a loud report, and soon afterwards a breadknife in a drawer inexplicably snapped into four parts, again with a sound like a pistol shot. In both cases he and his mother had been present. The four pieces of the knife are still in the possession of the Jung family.[4]

The dissertation is of particular interest because it contains the germs of some of Jung's later basic concepts. The medium in trance communicated words of 'personalities' whom he explained as personifications of unconscious and autonomous psychic elements. This indicated not only an intrinsic 'plurality' of the psyche; it also anticipated the concept of the so-called 'autonomous complexes' in the unconscious, which Jung later recognized as most important factors in psychic processes. The other subject of basic importance was a compensatory relation between consciousness and the unconscious: Jung observed that the personified complexes which manifested themselves in trances compensated for the conscious attitude of the medium and aimed at a greater com-

[2] In the 'Festschrift', *Die Kulturelle Bedeutung der Komplexen Psychologie*, edited by the Psychological Club of Zürich, 1935.
[3] *Collected Works*, Vol. 1 (New York: Bollingen Series; London: Routledge).
[4] Cf. for this and other parapsychological experiences: *Memories, Dreams, Reflections by C. G. Jung*. Recorded and edited by Aniela Jaffé. (New York: Pantheon Books; London: Routledge.)

pleteness of her character. Above all, it was the regular 'appearance' of a noble and aristocratic woman who, as a kind of unconscious ideal, compensated for the obviously too simple and still unformed character of the fifteen-year-old girl.

After a period of intense collaboration the 'somnambulistic' abilities of the medium declined, and she tried to make up for the fruitlessness of the sessions by fraud. At this point Jung broke off the seances. In the medium's later life, however, that 'higher personality' actually was established: the unstable young girl turned into an independent personality. She found a vocation which allowed her to develop her artistic abilities and won considerable distinction.

So-called somnambulistic phenomena seem most frequent during puberty. Jung offered the hypothesis that they represent attempts at character development. In his dissertation he wrote:

It is, therefore, conceivable that the phenomena of dual consciousness are simply new character formations, or attempts of the future personality to break through, and that in consequence of special difficulties (unfavourable circumstances, psychopathic disposition of the nervous system, etc.) they get bound up with peculiar disturbances of consciousness. In view of the difficulties that oppose the future character, the somnambulisms sometimes have an eminently teleological significance, in that they give the individual, who would otherwise inevitably succumb, the means of victory.[5]

Thus Jung attributed to 'somnambulisms' the same meaning that he later ascribed to the neuroses: he detached them from the causal viewpoint and investigated their final significance within an individual process of development.[6]

Although Jung's active interest in parapsychology never ceased, he did no further research work until his lecture on the belief in spirits delivered in London, 1919, to the Society for Psychical Research.[7] In this paper he explains 'spirits' and other

[5] *Collected Works*, Vol. 1, p. 79.
[6] Students of Jungian psychology will be interested to know that Jung's dissertation on the paranormal contains the first reproduction of a 'mandala' drawn in state of trance and representing a Gnostic system of the cosmos and its energies. It formed the culmination of the medium's manifestations, just before the sessions became increasingly shallow and meaningless.
[7] *The Psychological Foundations of Belief in Spirits*, *Collected Works*, Vol. 8. Cf. also *Proceedings of the Society for Psychical Research*, Vol. XXXI, pp. 75 ff.

such phenomena psychologically as 'unconscious autonomous complexes which appear as projections', or in other words as 'exteriorized effects of unconscious complexes'—thus taking up the arguments of his dissertation:

> I for one am certainly convinced that they are exteriorizations. I have repeatedly observed the telepathic effects of unconscious complexes, and also a number of parapsychic phenomena, but in all this I see no proof whatever of the existence of real spirits, and until such proof is forthcoming I must regard this whole territory as an appendix of psychology.[8]

When a new edition appears in 1947, almost thirty years later, Jung added a footnote to this sentence, based on his later concepts of the collective unconscious and of archetypes, which expressed a different viewpoint:

> After collecting psychological experiences from many people and many countries for fifty years, I no longer feel as certain as I did in 1919, when I wrote this sentence. To put it bluntly, I doubt whether an exclusively psychological approach can do justice to the phenomena in question. Not only the findings of parapsychology, but my own theoretical reflections outlined in 'On the Nature of the Psyche'[9] have led me to certain postulates which touch on the realms of nuclear physics and the conception of the space-time continuum. This opens up the whole question of the transpsychic reality immediately underlying the psyche.[10]

The theoretical reflections to which Jung here alludes will be discussed below in greater detail. But we may anticipate to this extent: Jung's investigations had led him to the conclusion that beyond the world of the psyche with its causal manifestations in time and space there must lie a 'transpsychic reality' where time and space are no longer of absolute but of relative validity: what the psyche experiences as past, present and future merges 'there' into an unknowable unity of timelessness, and what appears to consciousness as near and far combines 'there' into a likewise unknowable spacelessness. This 'transpsychic reality' is carried by the archetypes. From 1946 on Jung characterized these dominants of the transpsychic unconscious as 'psychoid' which indicates another merging, namely, that of psyche and matter.[11] This is an

---

[8] *Collected Works*, Vol. 8, p. 318.   [9] *ibid.*, pp. 159 ff.   [10] *ibid.*, p. 318.
[11] Jung distinguishes between archetypal contents and the 'archetype as such' (archetype in its strictest sense). The former appear as analogous motifs

obvious paradox which, however, is no more baffling than the familiar paradox of light which must be explained under some conditions as wave-like and under others as corpuscular. The psychoid (psychic-material) nature of the archetype, as well as the relativity of time and space in the transpsychic reality of the unconscious, became key factors in Jung's investigations of parapsychological phenomena and their characteristic incompatibility with the laws of time, space and causality. It is known that—because of elements of discontinuity in the atomic process—physics was faced with the same problem of non-causality. It will be shown later that parapsychological phenomena represent a kind of bridge between physics and psychology.

The postulate of an imperceivable psychoid background world colours the initial problem of 'ghosts' only to this extent: Jung could no longer maintain with assurance that these apparitions are projections of psychic complexes. What spirits really are, where they come from, why and where they are seen, remained for him an open question—and to this day science has given no conclusive answer.[12] Jung expressed himself very cautiously in his preface to the German edition of Stewart Edward White's *The Unobstructed Universe*:

> Although on the one hand our critical arguments throw doubt on every single case [i.e. of apparitions], there is on the other hand not a single argument that could disprove the existence of ghosts. In this regard, therefore, we must probably content ourselves with a *non liquet*.[13]

In Fanny Moser's book *Spuk*,[14] Jung describes his own encounter

[12] Cf. A. Jaffé, *Apparitions and Precognition* (New York: University Books, 1963). [13] *Das Uneingeschränkte Weltall* (Zürich, 1948).
[14] Baden bei Zürich: GYR. Verlag, 1950.

---

in myths, fairy-tales, dreams, obsessions, etc., throughout the world and throughout history. The 'archetype as such' is an unknowable (because unconscious) element which, however, has an organizing or formative effect on the so-called archetypal motifs. Moreover, it is the unconscious structural element, a 'pattern of behaviour', in such recurrent situations as birth, death, disease, transformation, love, etc., as well as in typical relationships such as mother–child, teacher–pupil, etc. Jung occasionally compared the 'archetype as such' with the axial system of a crystal which, as it were, pre-forms the crystalline structure in the mother liquid, although it has no material existence of its own. While the archetypal motifs are of psychic nature, the 'archetype as such' is psychoid.

with a phantom in England in 1920. He spent several week-ends in a friend's recently rented country house. During the nights he experienced various, increasingly violent phenomena like knocking, evil smells, sounds of rustling, and dripping. They aroused in him feelings of oppression and the sensation of a growing rigidity, and finally culminated in the apparition, or the vision, of a solid-looking half of a woman's head that lay on the pillow about sixteen inches from his own head. One eye was wide open and staring at him. The head vanished when Jung lit a candle. He spent the rest of the night sitting in an easy chair. Jung and his friend later learned what was already known to the whole village: that the house was 'haunted' and that all tenants were driven away in a very short time.

Jung interpreted some details of his experience as exteriorizations of psychic contents of the unconscious. But what remained an insoluble puzzle was the fact that the haunting took place in a special room of the particular house while during the week-days he spent in London, he slept splendidly in spite of a heavy working schedule. Thus this was a typical case of localized haunting ('ortsgebundener Spuk') for which to this day no sufficient scientific explanation has been found. The house was torn down shortly after Jung's visit.

In the beginning of the twenties Jung, together with Graf A. Schrenck-Notzing and Professor E. Bleuler, carried out a series of experiments with the Austrian medium Rudi Schneider, and witnessed materializations, psychokinetic, and other phenomena. Jung repeated such experiments in the thirties, again with Professor Bleuler and others. The medium was publicly known as O. Schl.; Jung never used these experiments for scientific publications. In the foreword to Stewart Edward White's book he summed up his impressions:

> Although I have never done any notable original research in this field [materializations, etc.] I do not hesitate to state that I have observed a sufficient number of such phenomena to be completely convinced of their reality. They are inexplicable to me, and I cannot therefore decide in favour of any of the usual interpretations.

The space-timeless realm of 'transpsychic reality' naturally tempts one to a variety of considerations and hypotheses not only about spirits but also about a 'beyond' and a life after death. Jung, personally, held the opinion that man would miss something essential if he did not indulge in phantasies and thoughts on this matter. His life would be poorer, his old age perhaps more anxious, but he would also break with spiritual and religious traditions that began at the dawn of human culture; and the disregard of such traditional roots generally exacts a heavy price. Time and again Jung was confronted with these problems when he had to deal with the mental suffering of elderly patients. If, however, man forms opinions about death and 'the beyond' he should keep in mind that in doing so he is entering the world of myths. They might have a healing effect and be beneficent but have nothing to do with science, or at any rate—as myth often forms the beginning of science—do not *yet* have anything to do with science. In his *Memories* Jung devoted the chapter 'On Life after Death' to his 'phantasy thoughts' or 'myths'. In his scientific work (his essay 'The Soul and Death')[15] he stresses the fact that the soul extends into a spaceless and timeless sphere. This offers much room for speculation but does not permit any final conclusions. A letter of May 1960 reflects Jung's opinion: in so far as the psyche is capable of telepathic and precognitive perceptions, he writes, it exists in an 'extra-space-time continuum'; hence the possibility of authentic postmortem phenomena:

> The relative rarity of such phenomena suggests that forms of existence within and outside of time are so sharply separated that crossing this boundary offers greatest difficulties. But this does not exclude the possibility that there is an existence outside of time which parallels the existence within time. Yes, we ourselves may simultaneously exist in both worlds, and occasionally we do have an intimation of the twofold existence. However, what exists outside of time is, according to our understanding, unchangeable. It possesses relative eternity.

In his parapsychological researches Jung devoted himself less to the problems of spiritualistic phenomena than to certain causally inexplicable events generally summed up under the head of *extra-sensory perceptions* (ESP). Literature and traditions of all countries and ages report examples of these occurrences, such as premonitions, prophetic dreams, inspirations, telepathic experiences, etc. Generally an unknown fact inaccessible to the organs of sense is perceived as an inner psychic image. It does not matter

[15] *Collected Works*, Vol. 8.

whether the fact actually has taken place in the past, is taking place now, or will take place in the future; nor is it of any importance whether it happens in the vicinity or in some remote part of the globe. *It is perceived here and now.* Usually these strange coincidences between inward image and external event are looked upon as mere chance occurrences. But that does not explain them. Scientific explanation, too, is doomed to failure as long as it is based on causality. How can a future event bring about a dream taking place now? How can a man dying in New York cause a clock in Manchester to stop, or a glass to shatter?[16]

Modern physics has recognized that certain laws of nature cannot be ascribed absolute validity. Rather, they are statistical statements of probability which hold only for the realm of 'classical' mechanics. In the realms of the very large (the cosmic) and the very small (the subatomic) exceptions to causality must be assumed. Similar exceptions may apply to psychic processes taking place in the borderline realm between consciousness and the unconscious. Here 'the connection of events may in certain circumstances be other than causal, and requires another principle of explanation'.[17]

Jung called this 'other principle of explanation' *synchronicity*. He defined it as 'a coincidence in time of two or more causally unrelated events which have the same or similar meaning'.[18] Thus synchronicity is an empirical concept which 'stipulates the existence of an intellectually necessary principle which could be added as a fourth to the recognized triad of space, time and causality'.[19] Jung stresses, however, that the concept of synchronicity should only be applied when a causal explanation is unthinkable. For 'whenever a cause is even remotely thinkable, synchronicity becomes an exceedingly doubtful proposition'.[20]

The 'coincidence in time' mentioned in Jung's definition does not mean an astronomical simultaneity dependent on clock-time.

---

[16] The necessity of dispensing with causal thinking is to this day the main reason for the objections raised against parapsychology as a science. The *Encyclopaedia Britannica* (1961) writes: 'The chief obstacle to a more widespread scientific acceptance of the findings of parapsychology, as some of the fairest and most competent sceptics have pointed out, is the almost complete lack of any plausible theoretical account as to the underlying causal processes.'

[17] *Collected Works*, Vol. 8, p. 421.
[18] *ibid.*, p. 511.   [19] *ibid.*, p. 511.   [20] *ibid.*, p. 461.

It refers to the inner image (dream, premonition, 'hunch', etc.) by which the past, present or future real fact is experienced actually present, here and now. Therefore Jung uses the term 'synchronistic' rather than 'synchronous'.

Along with the factor of time, the 'meaning' of the events is decisive. Synchronistic phenomena link facts that are not causally related and combine them into a meaningful experience. Jung offers the example of the sudden appearance of an insect resembling a scarab at the very moment when a patient was telling him the dream of a golden scarab. Here two facts (the dream, the appearance of the beetle) with two distinct chains of causality, are connected in one and the same meaningful experience. Prophetic dreams, premonitions, and similar happenings are likewise based on causes independent of those leading to the perceived event. In spite of their mutual independence the psychic image and the external event are linked together and the connecting factor is meaning.

Such meaning is not always expressed by the similarity or identity of content (real scarab, dream scarab) nor by the more or less photographic repetition or anticipation of an external event by an inner image. Frequently it is expressed in a symbol. This is true, for example, of those strange parapsychological happenings in which an object plays a part: it is meaningful that a clock suddenly stops, a glass shatters, a mirror breaks, a door opens by itself to 'announce' the death of a human being; for all these events may be interpreted as symbols of death or transition. Sometimes the anticipation of death is hidden in a strange archetypal dream-symbolism, as, for instance, in a dream series of an eight-year-old girl who died a year afterwards. Jung discussed it in his last essay.[21] The child had written twelve subsequent dreams in a notebook which she gave to her father at Christmas. In major archetypal images they represent impersonal, philosophical, and religious problems which far exceeded her understanding. The symbolically expressed meaning often places the foreseen event in wide,

[21] 'Approaching the Unconscious' in *Man and His Symbols* (London: Aldus, 1964), p. 70 f.—A dream from this series runs: Once upon a time I saw in my dream an animal that had lots of horns. It spitted other small animals on them. It twisted like a snake and lived like that. Then came a blue mist from all four corners and it stopped eating. Then God came, but there were really four Gods in the four corners. Then the animal died and all the eaten-up animals came out again alive.

impersonal, or religious connections. The myth-building and religious faculty of the psyche is here at work.

The meaning manifested in acausal synchronistic phenomena prompted Jung to postulate a transcendental 'knowledge' independent of human will and consciousness. He called it an *'a priori,* causally inexplicable knowledge of a situation which at the time is unknowable'.[22] Generally he used the term 'absolute knowledge in the unconscious'. Such a postulate goes beyond the boundaries of psychology and is of general philosophical importance. Therefore it seems all the more significant that the great physicist Wolfgang Pauli formulated a corresponding hypothesis, namely, 'the postulate of an order of the cosmos independent of our arbitrary will'. It is a metaphysical or an 'absolute' order to which 'both the psyche of the knower and the knowledge [based on observation of physical facts] are subject.'[23] In synchronistic phenomena this cosmic order of knowledge, or in Jung's phrase: an 'absolute knowledge'—both formulations embracing the soul and the universe—become manifest. The inkling of a universal metaphysical background independent of our arbitrary will explains the feeling of awe-inspiring numinosity that usually accompanies the experience of synchronistic phenomena.

It was characteristic of Jung's restraint in scientific matters that he waited more than twenty years before he presented to the public his decisive paper on synchronicity in 1952. The essay, *Synchronicity: An Acausal Connecting Principle* first appeared in German in the volume *Naturerklärung und Psyche,* together with a paper by Wolfgang Pauli on *The Influence of Archetypal Ideas on the Formation of Kepler's Scientific Theories.* Jung had formulated the concept of synchronicity as early as 1930, and applied it for the first time when he attempted to explain the method of the *I Ching* (Book of Changes), a Chinese oracular work of the fourth millennium B.C. At the beginning of the twenties Jung had come across the *I Ching* in Legge's English translation. Greatly fascinated, he experimented with it. The positive and meaningful oracular answers that resulted for him and for his friends confronted him with the unsolved problem of so-called mantic methods, which include astrology, geomancy, tarot-cards, and others. In 1923 he met Richard Wilhelm who, in collaboration with his learned Chinese friend Lau

[22] *Collected Works,* Vol. 8, p. 447.
[23] Jung-Pauli, *Naturerklärung und Psyche* (Zürich, 1952), p. 112. f.

C. G. JUNG AND PARAPSYCHOLOGY

Nai Süan, had after ten years of work just completed a new translation of the *I Ching*, along with a commentary on the various oracles. In frequent discussions the two scholars exchanged ideas about the book and the interpretation of its hexagrams.[24] When Richard Wilhelm died in 1930, Jung summed up in his obituary his explanation of the mantic method:

> The science of the *I Ching* is based not upon the causal principle, but upon a principle hitherto unnamed—because it does not exist for the Western mind—which I have tentatively called the *synchronistic principle*. My work with the psychology of unconscious processes compelled me many years ago to look around for another explanatory principle, because the principle of causality seemed to me insufficient to explain certain remarkable phenomena of the psychology of the unconscious. For I found first that there are parallel psychological phenomena which simply cannot be related to one another causally, but must stand in some other connection. The essential manifestation of this connection was the fact of relative simultaneity; hence the expression 'synchronistic'.[25]

Jung explained the positive results of the *I Ching* oracles as synchronistic phenomena, that is as an 'unexpected parallelism of psychic and physical events'.[26] This parallelism is experienced as 'relative simultaneity'. A meaningful link exists between the future event and the oracular pronouncement which is indicated by a sequence of fifty yarrow stalks or of three coins thrown here and now. Likewise astrology, especially modern character horoscopy, may achieve success in interpretation and prediction on the basis of a 'parallelism of psychic and physical events', i.e. of acausal coincidences between the constellation of stars and internal or external human events. Reading the stars is perhaps the oldest 'science' of mankind. The ancient stellar myths must be regarded

---

[24] In memory of this meeting Jung wrote (1950): 'I owe to Wilhelm extremely valuable elucidations of the complicated problem of the *I Ching* and also of the practical evaluation of the results attained. . . . When Wilhelm was staying with me in Zürich at the time, I asked him to work out a hexagram on the state of our Psychological Club. The situation was familiar to me, but not at all to him. The diagnosis that resulted was startlingly correct, as was the prognosis, which described an event that occurred later, and that I myself had not foreseen. To me personally, however, this result was no longer so amazing, since I had earlier already had a number of remarkable experiences with the method' (*Gesammelte Werke*, Band XI, p. 634).
[25] In: *The Secret of the Golden Flower*.
[26] *Gesammelte Werke*, Band XI, p. 638.

as projections of inner images on to the skies.[27] Astrology deals as much with these projections as with the stars, for horoscopes—because of the precession of the equinoxes—no longer coincide with the astronomical position of the stars. After Jung had studied the researches of Max Knoll into chemical and physical effects of planets upon sun and earth,[28] he began to doubt the absolute validity of synchronicity in the field of astrology and to consider the possibility of a causal connection between planetary aspects and the psychophysical dispositions of man. Thus the mode of operation of the traditional astrological method still represents a problem that eludes final solution.

Jung's method of approaching parapsychological phenomena was to combine extensive studies of historical and modern literature with careful observation of the facts of individual cases. He used the statistically computed results of J. B. Rhine's well-known researches into extra-sensory perception. In astrology Jung not only carried out a statistical experiment but frequently called for statistical investigation. But, in general, statistics to Jung were rather secondary. In his preface to F. Moser's book *Spuk* he writes:

> It is true that with the help of statistical method one can prove the existence of such [synchronistic] effects with more than sufficient certainty, as Rhine and other researchers have done. But the individual nature of the more complex phenomena of this sort forbids application of the statistical viewpoint, as it is itself complementary to synchronicity and therefore destroys the latter phenomenon—for it can do no more than eliminate synchronicity as a probable chance. We are, therefore, in this respect entirely dependent upon the well-considered and well-certified individual case.

The question of the circumstances propitious to synchronistic phenomena offers special difficulties. It has been observed that they occur (in the form of prophetic dreams, premonitions, psychokinetic phenomena, etc.) with greater frequency within the sphere of archetypal events, such as death, imminent danger of

[27] Projections are not made but 'found', Jung writes in a letter, June 1960: 'We must keep in mind that we do not *create* projections; rather, they happen to us. This fact permits the conclusion that originally physical and particularly psychological insights may have been read in the stars; in other words, whatever is farthest is actually nearest. Somehow, as the Gnostic surmised, we have "collected" or "crystallized" ourselves from out of the cosmos.'

[28] Set forth in the lecture 'Wandlungen der Wissenschaft in unserer Zeit' in *Eranos Jahrbuch* 1951, Zürich, 1952.

death, crises, catastrophes, sickness, outbreak of mental disease, etc. Since in archetypal situations man usually reacts with strong emotions, it would seem as if the emotion itself favours such happenings. In fact, during or because of emotion, the threshold of consciousness is lowered; the unconscious and its contents—the archetypes—break into consciousness and may begin to prevail.[29] In other words: man falls into the realm of the relative space-timelessness of the unconscious so that synchronistic events may take place more easily than in a firm state of consciousness. For this reason paranormal phenomena play a much greater part in the lives of primitive men, with their still feebly developed consciousness, than they do in ours; and for this reason the considerable gift for ESP that has occasionally been observed in children diminishes when they grow up and their consciousness has become solidly established.[30]

In his *Memories* Jung records a synchronistic phenomenon that he himself experienced in a state of great emotion. In 1909 he met Freud in Vienna. He was interested to hear Freud's views on precognition and on parapsychology in general, and asked him what he thought of these matters. Freud rejected this entire subject. It was some years before he recognized the seriousness of parapsychology and acknowledged the existence of ESP. Jung writes:

> While Freud was going on this way, I had a curious sensation. It was as if my diaphragm was made of iron and was becoming red-hot—a glowing vault. And at that moment there was such a loud report in the bookcase, which stood right next to us, that we both started up in alarm, fearing the thing was going to topple over on us. I said to Freud: 'There, that is an example of a so-called catalytic exteriorization phenomenon.'
> 'Oh come,' he exclaimed. 'That is sheer bosh.'
> 'It is not,' I replied 'You are mistaken, Herr Professor. And to

---

[29] Albertus Magnus (1193–1280) considered an *excessus affectus* as the explanation of 'magical influence'. Genuine magic such as is practised by medicine men, as well as the so-called 'excursions of the soul', probably are based on a capacity to raise and intensify such emotion by free will. In the emotion the threshold of consciousness is lowered and the way is open to synchronistic phenomena (subjectively experienced and described as all sorts of 'magic').

[30] Cf. Louisa Rhine, *Hidden Channels of the Mind* (New York: Sloane, 1961), which contains a large number of excellent examples of spontaneous ESP.—The material of spontaneous ESP that I have discussed in *Apparitions and Precognition* provides impressive examples of synchronistic phenomena in the emotional situation of examinations.

prove my point I now predict that in a moment there will be another loud report!' Sure enough, no sooner had I said the words than the same detonation went off in the bookcase.

To this day I do not know what gave me this certainty. But I knew beyond all doubt that the report would come again. Freud only stared aghast at me. I do not know what was in his mind, or what his look meant. In any case, this incident aroused his mistrust of me, and I had the feeling that I had done something against him. I never afterwards discussed the incident with him.[31]

In many cases, however, emotion is connected with some future event, while the experience in which it is anticipated occurs in a relaxed mood. This fact throws light on the nature of synchronistic phenomena. That is, on closer examination it turns out that the emotion as such has to be understood as a secondary symptom only. It may be present in the person who perceives the event, or it may be connected with the happening that is extra-sensorially observed, or it may be missing altogether. What is essential for the occurrence of synchronistic phenomena is the constellation of an archetype, and emotion is a determinant factor only in so far as it brings the unconscious (and the archetype) to the foreground.

Synchronistic phenomena take place within the sphere of archetypal events. This observation inclines us—because we are used to thinking in causal categories—to mistake the archetype for their transcendental *cause*. Such a conclusion, however, rests upon the same error that leads primitive man to explain virtually all events of life as results of 'magic causality'.

In reality the archetype must be regarded as the 'organizer' of the phenomena. It does not bring about a definite 'knowledge' of a logically unknowable event, nor does it 'arrange' a hallucination intentionally, as it were. Rather we must say: the 'unexpected parallelism of psychic and physical events' which characterizes these happenings is the manifestation of the archetype itself: its paradoxical, psychoid nature is split and appears here as image (psychic) and there as exterior event, occasionally also as object (physical, material). The archetype is arranging or manifesting itself in the facets of the synchronistic event. In other words: we are dealing here with an unusual manifestation of its 'emergence into consciousness' by means of a split. Such a split, as we know, underlies the process whenever something comes into conscious-

---

[31] *Memories, Dreams, Reflections*, London, p. 152; New York, p. 155 f.

ness: we recognize one thing because we distinguish it from another.[32] Under normal circumstances the split leading to consciousness takes place within the individual's psychic world, in his thoughts, dreams, intuitions, and experiences. In synchronistic phenomena, however, the various aspects of the archetype are torn apart. They present themselves at different times and at different places. This strange 'behaviour' may be explained by the fact that synchronistic phenomena are as much connected with consciousness as with the unconscious: the underlying or 'organizing' archetype is not yet fully conscious; it is partly still in the unconscious—hence the relativity of time and space. But partly it has penetrated into consciousness—hence the split of its psychoid nature into two or more distinct and perceivable facets.

In most cases it is not difficult to discover the particular archetype underlying or organizing the synchronistic phenomenon. It is most frequently death, the imminent danger of death, a disaster, etc. In the incident between Jung and Freud it was the impending end of a close human relationship of the archetypal father–son character. It may also be the expectation of a miraculous prophecy which plays a role in all mantic, or intuitive, methods. Jung considered the expectation of a miracle (the knowing of the unknowable) as the archetypal background of Rhine's ESP experiments. As a rule the scores of successes diminish when boredom sets in, that is, when the emotion connected with the expectation grows less intense and the archetype sinks back into the unconscious. Even in most cases of apparently unimpressive or banal synchronistic events the organizing archetype can be discovered. Louisa Rhine tells of a girl who anticipates that she will eat something uncooked, like spaghetti, and another girl unknown to her will say: 'That'll swell up in you'.[33] If we keep in mind the curiosity or even greed that prompted the girl to taste something straight out of a package of 'Lipton's chicken noodle soup', we recognize the emotional force of the impulse. The key words hinting at the archetypal background, however, are 'food', 'hunger', 'devouring'. These words refer to an archaic drive equal in importance to sexuality. Both are deeply rooted in the unconscious as instincts and play a part in the archetypal images of mythology and religion. They appear as gods of food, gods of love,

[32] We become conscious of death through life, of height through depth, of evil through good, and so on.   [33] *op. cit.*, pp. 221 f.

etc., and in this form they represent, as it were, the spiritual side of instinct. Both instinct and god-image are facets of one and the same archetype. Moreover, our instance contains an involuntary allusion to the old mythical tale or the archetypal representation of a woman's impregnation through eating ('that'll swell up in you').

So-called telepathy must also be understood as a synchronistic phenomenon. What happens here is the duplication of a psychic content—which appears both in the 'sender' and in the 'recipient' —rather than a parallelism of psychic and physical (or real) facts. From the psychological point of view, however, the question of sender and recipient recedes into the background: both are mere instruments of the autonomous archetype and of its 'orderedness' in space and time. Or else they may be understood as actors in the drama of an archetypal situation. Man, his conscious mind and will-power, are pushed into the background by the 'objective-psychical', the archetype. The impersonal, acausal process of 'arranging' (here the doubling of a thought content in two physically separate persons) may even occur if nothing is consciously 'sent'. If on occasion a thought seems to have been 'transmitted' deliberately it is not the intention that brings about the success but the emotional involvement of the individual or individuals—which in turn is connected with the archetypal character of the content or of the situation.

The well-known and often astonishing 'telepathy' between mother and child deserves special mention. The mother–child relationship represents an archetypal situation *par excellence*. For a long time after birth the two form a psychophysical unity, and a strong psychic relationship remains even longer. It has its roots as much in the unconscious as in consciousness. Another human relationship in which the unconscious bond is stronger than usual because it represents an archetypal situation is that between analyst and analysand. It rests upon the (one-sided or mutual) projection of unconscious contents.[34] In such relationships, based on an archetype and thus closer to the unconscious, it takes less to constellate a synchronistic phenomenon than with people who are not involved in an archetypal situation:[35] the knowledge of one

---

[34] Cf. C. A. Meier, 'Projektion, Uebertragung und Subjekt-Objektrelation in der Psychologie', in *Dialectica*, Paris, 1954.
[35] Cf. Celia Green, 'Analysis of Spontaneous Cases: 'Agent/Percipient Relationships', *Proc. S.P.R.*, Vol. 53, p. 181, November 1960, pp. 108 f.

partner concerning the thoughts and experiences of the other is more easily established because they may be drawn more readily into the relativity of time and space.

Synchronistic phenomena have their place between consciousness and the unconscious, between the knowable and the unknowable or between this earthly world and 'the transcendent psychophysical background'. Consciousness and this world represent, so to speak, the unfolding of entities and values which in the transcendental world must be thought of as a unity. Timelessness divides itself into past, present, and future, spacelessness into the various categories of space, and the unimaginable psychophysical (psychoid) unity of that realm appears as body and spirit or even as object and psyche. In the synchronistic phenomenon with its curious merging of time, space, object, and psyche, something of the original transcendental unity of the background becomes visible and can be experienced, and this is why man reacts to it with astonishment, with perplexity, or with awe. It is a natural paradox. The different aspects, or facets, originally united in the transcendental world, are not yet completely separated; they are not yet divided into single and unrelated entities. On the contrary, the physical and the psychic speak the same language. They express the archetype, their common but unrecognizable background. What binds them together is, as we have mentioned, no law of logic, no physical or 'magic' causality; it is an 'inner logic', namely, meaning.

The fact that synchronistic events are closely linked with the unconscious realm of the psyche, more specifically with the archetype, explains their unpredictable nature. *The contents of the unconscious operate autonomously.* This fact is one of the most important discoveries made by analytical psychology in the last decades. The autonomy of the unconscious contents makes all manifestations of the unconscious, including synchronistic phenomena, appear irregular. Regularity and predictability of events are guaranteed only where the concepts of space, time, and causality have absolute validity. But this, as we pointed out at the start, is no more the case in the border area between consciousness and the unconscious than it is in the realm of atomic or cosmic magnitudes. By their nonconformity with the law or causality, synchronistic phenomena remain exceptional, irregular, unpredictable happenings.

To summarize, we may say that synchronistic phenomena point to a background reality which is by definition unknowable, i.e. the unconscious. An unknowable background underlies physical phenomena as well. It is, as W. Pauli formulated it, a metaphysical order of the cosmos. The transpsychic as well as the transphysical sphere withdraw from direct observation and function autonomously; they are independent of our arbitrary will. After a careful evaluation of present-day science, especially of physics and the psychology of the unconscious, Jung concluded that the two background realities were possibly one and the same entity. This would be, he writes:

> as much physical as psychic and therefore neither, but rather a third thing, a neutral nature which can at most be grasped in hints since in essence it is transcendental.[36]

Such a transcendental unity would bridge the seeming incommensurability of the physical and the psychic worlds. Synchronistic phenomena point to that unity of 'neutral' or 'psychoid' nature; they are therefore not only of psychological but of general scientific importance.

In the work of his old age, *Mysterium Coniunctionis* (1956) Jung confronts causalism with an '*a priori* aspect of unity':

> The causalism that underlies our scientific view of the world breaks everything down into individual processes which it punctiliously tries to isolate from all other parallel processes. This tendency is absolutely necessary if we are to gain reliable knowledge of the world, but philosophically it has the disadvantage of breaking up, or obscuring, the universal interrelationship of events so that a recognition of the greater relationship, i.e. of the unity of the world, becomes more and more difficult. Everything that happens, however, happens in the same 'one world' and is a part of it. For this reason events must possess an *a priori* aspect of unity . . .[37]

The principle of synchronicity brings the long-lost unity of the world again within the reach of modern thinking. It 'suggests that there is an inter-connection or unity of causally unrelated events, and thus postulates a unitary aspect of being . . .'[38] From an historical point of view synchronicity, the 'acausal connecting principle' seems to be a compensating element in the disunion and dichotomies of our time. It is both of scientific and philosophical importance.

[36] *Collected Works*, Vol. 14, p. 538.   [37] *ibid.*, p. 464.   [38] *ibid.*, pp. 464 f.

XIII

ANTHROPOLOGY AND ESP

Francis Huxley

Social anthropologists study a number of things which should be of interest to parapsychologists, such as witchcraft, magic, divination, and shamanism. So far, however, the evidence they have gathered in these subjects has ignored the possibility of ESP rather than confirmed or denied it. There seem to be three reasons why this is so: firstly, that they, like the rest of us, tend to think in terms of 'body' and 'mind', without knowing how to bring these two categories together; secondly, that even if an anthropologist were interested in the study of ESP, the proof of its existence is so tricky that he prefers not to waste time on it; thirdly, that he is more interested in social regularities than in individual happenings. Moreover, though he may hear numerous anecdotes which tell him that the people he is studying take ESP for granted, he usually finds that the pronouncements of diviners are based on information previously gathered by ordinary methods, that they are of such general application that ESP can be ruled out, or that successes can be attributed to sharpened psychological insight gained through long practice.

The difficulty of the subject, distinct professional interests, contradictory evidence, and also, one must confess, a sense of embarrassment about ESP, all tend to one general anthropological conclusion: that tribal diviners (for instance) use quite ordinary methods to achieve their ends, and that their powers are of small calibre. It is true that on occasion the anthropologist hears of individuals renowned for their skill in healing or their oracular pronouncements, but these are as rare in their own countries as they are in the West and are not often met with in the flesh.

My own anthropological researches in Haiti are a case in point. The traditional goings-on there, which range from highly serious religious practices to vindictive essays in black magic, ought to have provided me with countless opportunities to witness telepathy, or bilocation, or oracular clairvoyance. In the end, however, I witnessed little that I found needed explanation in terms of ESP, though I made up for this by listening to amazing stories which were impossible to check. It was plain that Haitians believed in the existence of ESP—as do I—and delighted in this belief: but it was also plain that the evidence, such as it was, went the other way.

To speculate on ESP in such circumstances must appear disingenuous, and I can only justify what I am about to say by an indirect argument. It goes as follows: Wherever one finds divination practised, one finds a belief in the possibility of ESP: one also finds traditional techniques by which divinatory powers are said to be gained. Though it is possible to maintain that the end result of these techniques still has nothing to do with ESP, we cannot pretend that they do not bring about a change in those who practise them. Moreover, we find that very similar techniques are used throughout the world wherever there is a belief that man can enter into contact with the world of spirits, and with their help influence material events. It is therefore worth looking at the process which turns 'un profan', as they say in Haiti, into one 'avec connaissance,' in order to understand the human preoccupation with the supernatural which so often underlies ESP; we may also light upon an experiential understanding of the universal contrast between body and mind, or body, soul, and spirit, and by rethinking these ancient categories remove some of the difficulties which, for current orthodoxies, are inherent in the concept of ESP.

The techniques in question have one fundamental characteristic: a profound dissociation has to be provoked, during which the normal connections between consciousness and physical activity are severed. In one form or another, dissociation seems to be essential for ESP to occur, since it allows hidden thoughts to surface and removes the bane of self-consciousness. Two main directions open at this point: the practitioner may either remain conscious and embark on supernatural journeys while his body remains immobile and without sensation, as in certain kinds of shamanism; or he may lose consciousness, his body being then possessed by a spirit or a god as in cults such as voodoo or certain kinds of mediumship. (When trance becomes unfashionable—as it now seems to be in the West—mediums often practise a light dissociation without loss of consciousness, a state in which they are able to talk to, or to see persons present to their imaginations.) Both these processes show a number of interesting similarities, which make me think they are based on the same general principles. But as it is in voodoo that I have observed what happens at first hand, I shall focus upon it and describe its techniques for causing possession to follow on dissociation.

Voodoo is both a familial ancestor cult with private ceremonies, and a public cult during which the gods of an originally well-defined pantheon are invoked to possess their servitors. The ceremonies themselves are varied and complicated, as is the society and psychology which support them: all I wish to do here is to point to some more or less constant events, symbols, and activities in order to disentangle a basic process which often seems to bring ESP in its train.

It is important to remember that this process is at once psychological and social. The distinction between the individual and society is, philosophically, difficult to make and harder to maintain: society is inherent in individuals whenever they communicate, and if society were not to some degree a projection of individual activity it would not exist. We are faced with a similar but more acute paradox when it is suggested that something can be communicated between people without there being a transmission of that communication: persons and communications are here as difficult to distinguish as are individuals and society. If such is the case—which is by no means certain, for it is always possible that we may discover new modes of energy to explain ESP—we would be left with the old magical axiom that like influences like because of likeness.

This axiom is one that anthropologists must always make use of in understanding what is called primitive thought—which is not so different, come to that, from our own. The fact of likeness is held in the mind by means of a symbol or a ritual act, which draws its meaning from a number of different sources. Rationally speaking, the immediate result is confusing, as when a tree used as a symbol refers simultaneously to the original matter of man—we may remember the Portuguese word madeira, wood, and its etymological cognates matter, matrix, and mother—to its position as the central support of the universe, to the fact that it is the haunt of a cannibal spirit, to its use as firewood, or in leaf form as medicine. If its sap is white, this may refer to ideas about milk and, by extension, to the maternal line of descent; if it is a parisitic fig, it may symbolize the triumph of spirit over matter. The various allusions found in one symbol may refer to events on a physiological, cosmological, religious, or social level, yet when used correctly this profusion of meaning, far from clouding the issue, clarifies the process through which man creates himself in his world. It appears to be central to the problem of consciousness.

Though this is plain enough, and though Jung has charted a number of symbols which appear so often throughout the world that he calls them archetypes, the mechanism of the symbolic imagination still appears mysterious, and its choice of one image rather than another haphazard. Plainly, like does seek out like, but we must put some limits to this axiom to protect ourselves from the ambiguities of magical thinking. What I intend to do, therefore, is to show that man can provide a likeness to the outside world in the experience of his own body, from which he creates his archetypes. We can immediately see how this reflection of the outer on the inner can become distorted: the simile cannot be much more than a makeshift in rational terms, since personifying thought and inanimate process are in some ways irreconcilable; besides this, the ways in which we are taught to become persons bring in their own complications in the form of neuroses and psychoses, from which an individual must escape before he can fully see the likeness within him of the outer world. We can follow the main lines of this process in Haitian voodoo, starting with the symbol of a tree which there occurs both spontaneously in dreams and in the visions accompanying individual neuroses, and traditionally in rites of initiation.

There is a sense in which we can say that in Haiti the course of mental breakdown and its cure on one hand, and the ritual of voodoo on the other, are reflections of the same process, and that the parallels they show are more than coincidence. Some of these similarities, briefly, are as follows. Disquieting dreams before mental breakdowns frequently concern snakes, while visions during breakdowns themselves are often of a god called Grand

Bois, Great Tree. Grand Bois gibbers ferociously, cannot speak, and prowls the woods like a wild animal. But he is the lord of the woods and of herbal medicine, and often helps the afflicted visionary by giving him a recipe for his cure and a method by which spirits can be invoked in the future for oracular purposes. In his other, more benevolent aspect Grand Bois is called Legba, being equated with the central post of the temple dancing floor, with the euphorbias planted around houses to ward off evil spirits, and with the gate leading from this world to the next. In his primitive form he manifests during initiation ceremonies at the time when the novice priest is made to stand with his back against the centre post, while an egg is broken near his head. It is down the centre post that the gods are thought to come by a serpent track when possessing their servitors, while eggs are the food of the serpent god Damballah who likes to coil about trees. The import of this action is to turn a man into a serpent-entwined tree haunted by the gods: an action performed by ritual means. The appearance of snakes and Grand Bois in neurotic circumstances shows that the process is ripe to take place, but that without conscious direction, or help from the vision, the forces involved end in a breakdown.

Mental disorders in Haiti are thought to be caused by a rush of blood to the head which unbalances the soul: the blood is persuaded to return to its proper place, and the soul restored to its balance, by a head-washing rite. The same rite, making use of the same medicinal and magical leaves, occurs during initiation in order to establish the capacity to be possessed without going mad.

A universal characteristic of initiation is that it symbolizes a mock death and rebirth. In Haiti the novices are said to be suffering from a great disease, of which they are cured by initiation: a disease which the frequent cries of 'there is malice, o!' suggest is the ritual enactment of secular nervous disorders, so often caused by suspicion, malice, frustration and fear.

Similarities like these show that ritual can make use of a process which, when it occurs spontaneously and without direction often ends in breakdown. We must not say that such a cult is a form of neurosis itself: what seems to be happening is that it uses the forces inherent in neurosis and directs them to definite ends, breaking down certain patterns of thought and behaviour and replacing them with new ones.

The mechanism behind spontaneous and ritual breakdown and cure is the same. In both, the breakdown takes place after a long period of incubation: the spontaneous breakdown when social pressures, frustration, anger, suspicion, and hatred reach unbearable proportions, to be triggered off by often quite trivial incidents; the ritual one as a result of what can be understood as a number of powerful suggestions and commands hidden in ritual and symbolism. Since these are traditional and therefore constant, it is convenient to focus on them rather than on individual neurotic episodes, and to point to the two main events which are consciously brought about during voodoo: dissociation and possession.

The preparations for a voodoo ceremony are lengthy, consisting of designs drawn in flour or ash upon the ground, libations poured at various points, prayers and songs to Christian saints as well as to the gods of voodoo. When they are accomplished, the drums begin to sound and the choir of men and women attached to the temple sing and dance in praise of the gods. The mounting rhythm of drum and song, and the continual effort of the dance, lead to the first signs of the gods' presence among the dancers: a certain abstraction. It is the drummers who largely provoke dissociation: they are skilful in reading the signs, and by quickening, altering, or breaking their rhythm they can usually force the crisis on those who are ready for it. The dancer thus singled out falters, feeling a heavy weight on neck and in legs, while a darkness invades his sight and mind; he loses his balance and totters with great strides from side to side of the dancing floor, a bewildered or agonized expression on his face. He may collapse among the audience at the sides, who put him back on his feet and send him for another voyage over the floor till the buffets of sound have their full effect. Suddenly a new expression dawns on his face and he draws himself up in an attitude which is often instantly recognizable: the god has possessed him or, as the usual expression goes, he has been mounted by the god as a horse by its rider. The god then stalks about the floor, paying his respects to the priest, taking part in the ceremony if necessary, admonishing persons in the audience or giving them his blessing, till he slowly or suddenly absents himself. The dancer comes to himself with a bewildered expression and usually retires for a time to recover.

The sensations that occur during dissociation and possession are often unpleasant, though priests and priestesses, who may be possessed several times a day, slip in and out of this state easily enough. Besides the feeling of heaviness and loss of balance, there is fear and horror, nausea, vertigo, darkness, odd tinglings of the body. Haitians know a possession to be genuine by certain signs, principally that the god's 'horse' does not remember what he did when possessed—though this knowledge may come to him later in a dream. He may also become insensitive to pain or accomplish unusual feats, as when an arthritic woman, possessed by Damballah, swarms up a tree without difficulty. From time to time members of the audience, not initiated, may become possessed by a wild spirit, but more often they continue in the drunken wobble of dissociation and the crisis remains unresolved by the appearance of a god. (A person in the first stage of dissociation is in fact said to be 'saoulé', drunk by the nearness of a god.)

These experiences are of much interest, for they exhibit the same range that occurs in disturbances of the inner ear. The inner ear modulates postural attitude, muscle tonus, breathing rhythms, heartbeat, and blood pressure, feelings of nausea and certain eye reflexes. It appears that the apparatus of drumming, dancing, and singing can not only affect this centre but is aimed at it in an effort to dissociate the waking consciousness from its organization in the body. This brings about that disorder of self-control which in ordinary life is called madness.

What can we say about this coincidence? Can we in some way relate the psychological structure of the social person to the workings of the inner ear? It is possible that we can if we are simple-minded enough to ask what relationship exists between psychological tension and tension of the musculature.

The concept of psychosomatic disease shows that certain physical illnesses are a manifestation of definite psychological difficulties, and they are classed under the rubric of tension. A simple example is that of arthritis, now being relieved by a medication that lowers muscle tonus. The way in which mind influences body in such diseases is often considered to be difficult to understand, largely because something in our experience has taught us that 'mind' and 'body' are categorically distinct. Physiology, for a time, made this kind of reasoning respectable by studying organic processes in isolation both from each other and from the purposeful activity of the organism. It is now coming to different conclusions. We know for a start that muscular activity is dependent on a large number of nerves, collected into different nerve bundles and conducted separately to the brain. Analysis of extensive muscular movements shows that they are organized by a definite part of the brain *in terms of the movement as a whole*, not as a serial and partial accommodation of individual muscle fibres to each other. We can train ourselves to twitch individual muscle fibres, but they are organized and moved as a complete gesture and not individually. The many odd facts found in hysterical or hypnotic anaesthesia shows that this is true: the area thus anaesthetized may be served by different nerve branches, some of which serve neighbouring areas which are yet fully sensitive. In some way, therefore, our muscular and sensory activity is organized by means of an image which overrides purely anatomical connections.

It would be surprising if matters ended here. Freud, in his essay on Wit and its relation to the Unconscious, remarks briefly on the role that movement plays in organizing many parts of our psychological life. In order to understand a gesture that one sees, he says, one perforce imitates it, and in so doing sets a standard for its meaning in one's feelings of bodily innervation. He continues:

> The recollection of this innervation expenditure will remain the essential part of the idea of this motion, and there will always be methods of thought in my psychic life in which the idea will be represented by nothing else but this expenditure. In other connections a substitute for this element may possibly be put in the form of other ideas, for instance the visual idea of the object in motion, or it may be put in the form of the word-idea; and in certain types of abstract thought a sign instead of the full content itself may suffice.

We can see here how meaning can be created by physical imitation of what is happening, and how this can become concentrated into a visual or verbal equivalent which stands for the entire situation. Work with hypnosis has come to a similar conclusion from the opposite direction. If we want to produce a change in the body of someone who has been hypnotized, it can only be done by suggesting a situation in which that change would be appropriate. Thus the heart can be made to beat faster only by inducing, say, a fearful apprehension, not by the direct command 'your heart will now beat faster'. The same is true for all vegetative responses which together make up what is truly unconscious within us, but can yet be called up by means of a conscious image.

It might also be possible to follow up Freud's line of thought in connection with the origin of language, perhaps the prime mark of human consciousness. Sir Richard Paget, harking back to Plato's Cratylus, in fact proposed that men first communicated by means of imitative gestures of the hands, and that these gestures extended to movements of the lips, jaws and tongue which unconsciously followed suit. The idea has been strangely neglected, but is by no means absurd: the speech centres in the brain seem to be developments of the hand-gesture centres, for instance, and the imitative nature of language would help account for such anthropological oddities as the belief in the magical power of names, and obsessions with spoken ritual, both of them attempts to influence things through their representations. Our non-magical habit of abstraction hides from us the fact that we become conscious of things both mentally and physically, and that the key to understanding the relation of mind and body lies in the images which represent this dual experience to us.

In any event, there are a number of slightly unorthodox Western therapies which work on the principle that neurotic tension is coterminous with muscular tension, and that neurosis can be relieved by relaxing the muscles and re-educating the general posture of the body. A start on these lines was made by Mathias Alexander, who found that by altering the set of head on neck and shoulders, and the posture of the body in sitting and walking, he could bring about fundamental changes in psychological attitudes.

'Attitude' is, etymologically, closely related to 'aptitude', and can be used to mean both a mental set and a physical posture. The kind of metaphor enshrined in this double usage should be attended to carefully: like a symbol it refers to two different rational categories, and it is difficult to know which of them should be regarded as primary. But the problem is no doubt a false one. The word can be used metaphorically because the same process underlies both usages: the body readies itself to meet a situation as soon as the mind is aware of it, and this readiness is the attitude in both its aspects. We can perhaps translate other metaphors in the same way, by noting that Alexander's interest in the back of the neck can be found in voodoo both directly and indirectly, in experience and in ritual. The direct experience is found among many voodoo servitors who say that the first throes of dissociation are felt at the back of the neck, in the form of a great weight; and those who practise oracular divination without being possessed often feel the god sitting on their neck while he whispers into one ear. Indirectly, we find that an important act during initiation is the stubbing out of a candle on to the nape of the neck, in order to seat the soul firmly in its place. The ritual thus directs the attention of the novice to the place where he will feel dissociation first happening, and so the metaphor can be seen to have physical as well as intellectual reference.

The circle formed by a symbol or metaphor that takes its appearance from the external world, the psychological meaning of this, and its physical analogue, can be shown to exist in other ways. In his work on dreams, Freud briefly mentions a parallel theory put forward by Scherner, who said that dreams tried to represent symbolically the nature of the bodily organ from which the dream-stimulus proceeded. This revival of dream-interpretation by means of symbolism was not followed up by Freud, who was more interested in analysing the psychic process in dreams, but he did not deny its general truth; and it has now been supported by the finding that muscle tonus increases during dreams, while the eyes flicker from side to side as though viewing a real landscape. It seems then that images are experienced in connection with the bodily activity which we would normally carry out when faced by the real thing, and that when such images are held in the mind, the body carries out the required movements on an almost imperceptible level—a fact which has lately been made use of in a novel system of exercise. In other words, a psychological attitude is coterminous with its physical one.

Why should this be? It seems that an intention to act creates in the body a readiness to carry out that action—it stretches towards its object, as the word properly implies—and it is possible to demonstrate that the attitudes in which we receive confidences from a friend, or insults from an enemy, in which we endure pain or abandon ourselves to pleasure, are subtle forms of body English: that is, they are demonstrations in the body of a meaning held by the mind. Attitudes are thus forms of communication, and the frame in which communications are understood: they are intentional.

When an intention is socially unacceptable it is said to be suppressed. If we look at what is happening on the plan of the body, the attitude which is disapproved of remains, but is stopped from being fully expressed by a counter-attitude. What we call 'tension' seems to be the sum of our habits, our intentions and the attitudes the body takes up to express them, and the counter-intentions and attitudes which are the expression of shame, guilt, fear, and anxiety. Society appears to repress or dissociate the simultaneity of intention and attitude: by making us live more reflectively, in our minds, it stops us being aware of our bodies which have to act out conflicting attitudes. When we act socially it is difficult for the mind to disentangle the different meanings of these superimposed attitudes, and it allows itself rather to be supported by the feeling of tension which they produce. However, if intentions belong to the acting body as much as to the mind conscious of intention, part of the mind will always be imprisoned in physical tensions where the body is made to contradict itself; and it remains imprisoned until the meaning of an attitude emerges into consciousness, distinct from the sensations it produces.

Freud has mapped several regions of the body where erotic intentions are imprisoned as psychological complexes, but we can extend his map to take in the body as a whole. His somewhat eccentric follower, Reich, invented the illuminating phrase 'character armour' to describe the muscular tensions, especially in hips and back, which he found to exist wherever sexuality was repressed; and he showed how this rigidity of the body acted to stop a number of natural, rhythmical movements by which intention could be consummated. Character armour, he showed, is the antagonist of action.

The process of dissociation attacks this character armour, undermining the conscious element of self-control and freeing the unconscious, spontaneous ones from its hold. The psychological ego, in which normal consciousness operates, must be located in this complex of mental and physical attitudes; and when dissociation happens as in voodoo, the ego is broken down together with its organization of the body. As self-consciousness disappears, so does the physical attitude through which a man keeps contact with the world he lives in—he loses his balance, his muscles no longer obey him, and the various relationships he has created between knowing, feeling and moving are destroyed.

If dissociation were not followed by possession in voodoo (or, in the arts, by inspiration and intuition)[1] it would be merely a useless and dangerous interlude. But how does possession take place? Looking at men and women who are possessed, an obvious fact strikes one: the gods who come all have ritualized attitudes and mask-like expressions. If the dancers have been made to lose their habitual attitudes, they have now taken on new ones, and it is these which the gods partly consist of.

An interesting study by Michel Leiris on the Ethiopians of Gondar suggests that these attitudes are consciously learnt. In the Ethiopian cult, where possession also takes place, novices are put through certain ritualized movements known as the *gurri* which appear to be representations of dissociation; they also learn the exact behaviour shown by the gods of the cult. By a process of play-acting, the internal meaning of these movements becomes real and suddenly overtakes them: the *gurri* becomes a real dissociation, the play-acting a real possession.

This learning process is not so marked in Haiti, where spontaneous dissociation and possession often take place. There is no doubt that the more a person becomes possessed, the more perfectly he may come to represent the various gods possessing him: practice indeed makes perfect. However, it is generally recognized that the most convincing characterizations are shown by those who have an innate disposition corresponding to one or other of the gods, and are not merely an acquired skill.

The main pantheon of Haitian gods manifests a series of basic human attitudes. There is warlike Ogoun, oracular Loco, the priest-like Legba, the bawdy corpse god Guédé, the sorrowful love goddess Erzulie, the magician Simbi, the wise serpent and first progenitor Damballah, and so on. Outside this pantheon there are unnumbered gods who have arisen out of special circumstances, either social or individual. An interesting group of these

[1] T. S. Eliot's remarks in *The Use of Poetry* (London: Faber, 1964) are of interest in this connection: 'I know, for instance, that some forms of ill-health, debility or anaemia, may (if other circumstances are propitious) produce an efflux of poetry . . . To me it seems that at these moments, which are characterized by the sudden lifting of the burden of anxiety and fear which presses upon our daily life so steadily that we are unaware of it, what happens is something *negative*: that is to say, not "inspiration" as we commonly think of it, but the breaking down of strong habitual barriers—which tend to re-form very quickly.'

gods is known as the Petro nation, who are ruthless, cruel, violent, and expert in magic. We can read this general character both as a legacy from the old days of slavery and as a reaction to the present misery of Haiti, damned as it is by poverty, over-population, a broken-down society, and a tyrannical and wasteful government. In the south where I worked, the Petro gods seemed to be as numerous as the other, more beneficent ones, and may have outnumbered them, and it was striking to see the attitudes they manifested. One might understand them in terms of all the wild intentions society carefully represses, while the beneficent gods represent the ideal persons of society, confident of their position and their lasting meaning.

At the time I unfortunately did not specifically note the variety of these different attitudes, having not yet understood their importance, but my general impression is that they exhibit three main tendencies. The gods of the pantheon, being gentle and good unless wilfully crossed, are statuesque, well controlled in their movements, their faces faintly hieratic—or, as in the case of the Guédés, showing considerable human depth even in their most theatrical moments. Some Petro gods are wild, furious, and ambitious, others stern, repressive and antagonizing. It is not always easy to distinguish these Petro attitudes, but I believe they reflect the two terms in one contradiction: the furious attitudes representing a primitive urge to break out of an impossible situation, the stern ones being a caricature of the attitudes which repress them. The gods of the main pantheon would then represent a stage beyond this contradiction, when the elements of the person are not at war together.

It is thus possible for the various characters of the gods to represent attitudes already stored up within the men and women possessed by them in the form of unacted intentions. The ritual during ceremonies largely guides the choice of which will emerge at specific times, while the carnival of essences and caricatures is directed and brought to a safe conclusion by the priest, who often has remarkable psychological insight.

There are several stages in voodoo, which may have some relationship to the kinds of gods which possess a man at any time. The first stage is called *hounsi bossal* and refers to a servitor who is uninitiated and can be dissociated but not possessed. The second stage is called *hounsi canzo*, meaning a man who has been initiated, has been visited by his 'head master' or patron god and possessed by him; the third stage is called 'the taking of the asson', the asson being the rattle which is the mark of a priest or priestess. Those initiated into the priesthood learn the direction of ceremonies, the secret and necessary passwords and gestures, the public songs and prayers which go with them; a knowledge of magic and herbalism, and the ability to call up ancestral spirits from the waters of death and make them speak. The fourth and apparently final stage is called 'la prise des yeux', the taking hold of the eyes. By this is meant an ability to be possessed without losing consciousness, to exercise certain occult powers such as telepathy and clairvoyance —bilocation is also mentioned occasionally—and to gain increased powers of healing and curing.

Words for consciousness, knowledge and wisdom are so often related to those meaning sight that the choice of the phrase 'la prise des yeux' for a highly active state of awareness may pass unnoticed. But why is this metaphor so frequently used unless it means what it says? If we take it literally we can hazard the guess that vision is deeply implicated in the state of self-consciousness, and that when the ego is dissociated from the visual experience, which includes non-visual sensations in the eyes, it has lost the light of consciousness and all then becomes dark. Evidence to support this position could be taken from Freud's findings that blindness is a symbolic castration, or in Schilder's analysis of the body image, which he concluded was made up of two parts, one of them organized around motor and tactile elements, the other around vision and intellection. The connection between vision and consciousness can also be seen in physiology, in the form of the reflex mediated by the inner ear between the action of the neck muscles in turning the head and the corresponding pattern in the eye muscles which keep the point of fixation from being lost. As the eyes co-operate with the semicircular canals in maintaining postural balance, the place of vision in the psychological organization of the body becomes apparent. If, then, the eyes can be freed from the neurotic patterns of muscle tension which extend through the entire body, this must mean that a higher stage of consciousness has been reached. I use the word 'higher' purposefully, to underline the fact that the metaphor has a physical reference, and that a consciousness of the muscular activity of the eyes when related to the fact of seeing will bring about a distinct experience of self-knowledge. Traditional descriptions of this experience can be found in works on yoga, which also makes use of the concept of ascent, from base of spine to top of head, as a metaphor of the development of consciousness. The penultimate stage of this development is located in a centre controlling the eyes. The physical analogue of height is thus paralleled by a psychological one, if the connection between seeing and knowing points to a real fact of experience. We could then say that 'la prise des yeux' is an understanding of that process by which seeing and knowing are normally conflated, and a liberation from its consequences.

A series of hypnotic experiments carried out by Dr. B. A. Aaronson brings this argument down to a more accessible plane, by showing that perception influences mood. He began by suggesting a number of different visual modes to his subjects, and found that when they experienced the size of things as being smaller than normal, they underwent a schizoid withdrawal. Increasing the size of things produced a feeling of panic. Clear vision was sometimes responded to with anxiety, but not always; blurred vision with a sense of 'belle indifference'. A sense of no depth produced fatigue, hostility, and depression as found in schizophrenia; expanded depth brought about a euphoric, creative state of mind, as though a hallucinogen had been taken. This last finding is of particular interest, since the expansion of consciousness under hallucinogens has often been likened to certain mystical states, in which an organized increase of perception goes hand in hand with a sense of all-knowledge and a unitary state of consciousness. At all events, the hypnotic experiments show how closely connected visual perception can be with mood—it is of interest that similar suggestions connected with hearing did not produce basic changes in character—and since mood and attitude are related, the results suggest that perception is altered at least partly by changing the tone of the muscles concerned with vision, which I imagine include those of the face and head as well as of the eyes.

My argument so far is that intention and attitude are two words for the same state, and that such a process as voodoo allows buried intentions to act themselves out and become conscious: first to other people, then to oneself during dreams, finally in full consciousness. As he progresses, the initiate gains a command of his own person. The warring intentions which at first were largely unconscious patterns of tension are resolved into easily recognizable figures, the gods, each with their specific character. They act themselves out theatrically in a complicated ceremonial till the initiate is able to know these patterns of activity with his mind and to accommodate them in the consciousness of himself. As he does so, his sensory awareness will become greater, and faculties dependent on the full experiencing of all that happens—such as ESP—will come to life.

The use of symbols in voodoo, as in yoga, allows us to see this process at work in various parts of the body when we take them literally. For Haitians, the conscious soul is located in the nape of the neck; oracular spirits or 'mystères' live in the belly, from which they speak ventriloquially; the gods and the spirit of man live in the head. Haitian psychology is based on this threefold schema, and through it we can understand why a rush of blood to the head is thought to cause both madness and dissociation, since it displaces the soul and infects the spirit with turbulent heat: the spirit, like the spirits of the dead, is properly at home only in cold water or a cool head. As the soul is displaced, so the characteristic features of dissociation appear, features which we can parallel in disturbances of the inner ear.

Other symbols and ritual acts seem to be transcriptions of the experience of dissociation and possession. The symbol of the tree is one such. A number of songs have to do with trees and the magical properties of leaves, which are guarded by Grand Bois, by Maman Travail whose home is the fig tree, by snakes and by the water god and magician Simbi. A central image is of the old fig tree, haunt of evil spirits, falling to earth where its leaves are eaten by goats; it is the symbol of unregenerate and uninitiated man, and of that complex of attitudes which falls apart at dissociation. The new attitude, representing possession by a god, is seen as a new tree, and is figured by the appearance of Grand Bois both in spontaneous mental crises and in ritual where the novice is made to stand with his back against the image of the World Tree, the centre post of the temple. Anatomically, the tree would represent the spinal column—a frequent identification throughout the world. The constant association of Snake with Tree seems to describe at least two physical experiences. One is the serpentine flexing of the spine which is often found in dissociation, and which in part has a sexual reference; the other has to do with spasms of the gullet connected both with difficulties in breathing and with feelings of nausea. Since snakes are the guardians of treasure in Haiti, and treasure psychoanalytically can be understood by way of the anal complex, it may well be that the symbol of the snake embodies sensations of the gut as a whole.

We find a similar coincidence of image and physical act in shamanism. The crises which initiate a novice are, according to the literature, fearful and agonizing: sensations of nausea, vertigo, formication, and an apparent disarticulation of all the joints are common, while the typical shamanistic ascent to heaven, which describes the liberation of the spirit from its prison in the body, is described either as a ride on horseback or as the climbing of the World Tree. This ascent has numerous stages, each with its own rewards, dangers, and powers: according to Mircea Eliade, it is to be seen as the source of Indian yoga, where the Tree is called Mount Meru, also a name for the spinal column; that which climbs is called Kundalini, the serpent power, the stages on the ascent being called chakras, centres of psycho-physical energy and control.

It thus seems that we are dealing with an archetypal situation, regularly transcribed by very similar images. These images are also to be found in our society. I have lately made some experiments with people who have taken hallucinogenic drugs, and asked them whether they can find a physical reference for each image which appears to them when their eyes are closed. One young woman found that the image of a tree could be located in sensations up and down her spine; she also experienced what is a very old Indo-European image of heaven and earth, that of the tree rooted in earth whose branches interlock with those of a tree growing upside down, its roots in heaven. The image of this second tree she found to be rooted in tension patterns in head and neck, branching down into the muscles of the shoulders. Images of snakes, of the phoenix, of fire, volcanoes, mountains, eggs, and lotus flowers in turn were all found to be references to definite sensations in various parts of the body.

Fantasies of other kinds can be understood in this way, for instance, that of rebirth. I choose this one because initiation is universally understood as the experience of a mock death followed by a rebirth. The psychological process has a close parallel in a number of hallucinogenic experiences which I have witnessed. One such again concerned a woman who felt that she was about to give birth to herself by being born. Nothing developed out of these feelings, which caused her much anxiety, till she was persuaded to act as though the fantasy were real. After some forty minutes of strenuous effort, during which she was told to breathe deeply and to flex back and hips, she encountered the typically unpleasant sensations found in voodooistic and shamanistic crises: nausea, fear, anxiety sometimes amounting to terror, pain, giddiness, and formication. The effects of overbreathing, which produces a mild form of tetany, and the euphoriant effects of the drug, combined to unlock the complex of tensions which had previously armoured her and which were obvious both in her musculature and in her psychological difficulties. She emerged from this armour, as it were, through two main centres of constriction: one at the top of her head, giving rise to the idea of being born; one between her legs, where the image of giving birth arose. The flexures of the body suggested to her were typically those found in crisis states, and allowed her to feel head and trunk through one rhythm rather than as a disconnected series of tension points. (The new and intensely enjoyable sensation of herself which resulted has its obvious parallel in yoga, where base of spine and top of head are regarded as being the first and last stages of spiritual development.) After this event, she enjoyed the typical experience of being liberated, of superior knowledge and of enlarged perception.

There are thus many indications that symbols refer to physical states and that traditional systems make use of the common basis of intention, attitude, and image in order to develop self-awareness. We can look further at this triad with the aid of more orthodox psychological thinking. Looking at the play of children, for instance, Piaget saw that it could be formally broken down into two components: an intellectual one, because children learn the nature of what they play with during play; and a physical or sensori-motor one, because in playing with a ball the child must learn how to judge distance and speed and to use his hands in catching and throwing. We can describe the first component of being concerned with the assimilation of reality through knowing, the second with the accommodation of the body to the things it wishes to know about. This division is formal rather than real, as Piaget recognized, and it is truer to say that for every act of assimilation there is one of accommodation and vice versa. We learn what a ball is by catching it, and the fact that a ball is something to catch teaches us the use of our bodies in catching it.

Some recent experiments have shown that cats come to know and recognize their surroundings when they are able to move about in them, but not when they are held immobile. This fact, I imagine, is generally true. Our senses are intimately connected with our motor activity, and the world we know is defined in terms of how we act towards it. It follows that the frame in which we put our knowledge of the world is the frame of the body itself.

For some time now psychologists have been interested in what they call the body image, which is an image of ourselves in terms of our surroundings and of what we feel capable of doing and feeling; it is often at variance with our real potentialities and our visual appearance. The body image reflects our habitual activity, and is the sum of the attitudes we take up towards other people, towards work, pleasure, food, space, time, and so on. (Its various functions bear an extraordinary resemblance to what Freud called the pre-conscious, which is the transforming mechanism in dreams and creative activity.) What we are conscious of in our activities is organized according to this body image: it frames the ideas we make, gives form to what we do and to the information brought to us by our senses. To have an image of something outside us we must also have an image of ourselves, and the body image will therefore be the frame in which all images concerning the outer world are received, organized, and understood.

This formulation may seem extreme, but it is unavoidable if we are to understand how the symbols in voodoo, shamanism, and hallucinogenic visions refer not only to appearances outside the body but also to sensations of the body itself. We are here at the crux of the problem: there seems to be no way in which we can distinguish emotionally between the images coming to us from the outer world and those transcribing experiences of ourselves. At a certain level, the tree that we see is the tree that we carry about within us as the spinal column; the snake that we fear is the place in our bodies where we feel fear as a snake, the candle and the flame is recognized by us in the place where our bodily processes are turned into conscious light. We do, in fact, continually see our own likeness in external appearances, and all traditional systems make use of the axiom that like influences like, by making their initiates enter physically into the meaning of those symbols and metaphors which transcribe subjective experience in terms of natural events.

The capacity of the body image to contain everything has its own range of symbols, sometimes plainly evident as in the cabbalistic figure of Adam Cadmon or in creator gods whose bodies turn into sky and earth, hills, forests, and rivers; more obscurely in sea monsters and dragons, or diagrammatically in houses, palaces, and temples. In such images there is room not only for he who experiences them and for every part of his experience but also for all others who experience the same thing: the All is included by the One, and the One by the All. It may be that the state of omniscience reported in various mystical writings, which one could say was ESP writ large, stems from participating in such an image. We have little enough evidence to work on at this level, and it is a question whether such findings as those of the Roses among the Australian Aborigines can be taken in this way or not. They reported that among these people the awareness of the death of a relative sometimes arose telepathically on seeing the bird or animal which was the totem of the dead man. The totem stands for the ancestral principle, among other things, and thus includes all those descended from it: its use as a portent concerning one of these descendants is therefore understandable, though how the observer knows which one is meant remains mysterious.

It has long been recognized that symbolic thought is basic to mankind, and that traditional cultures classify the world through its logic. In doing so, they form themselves in the image of what they see, and of the processes underlying that: and if this image is clear in themselves and experienced in their person as an act of what we may call the physical imagination, they are in that state which Levy Bruhl, somewhat pejoratively, called 'la participation mystique'. (Gilbert Murray's description of how he felt towards the people with whom he was in the process of exercising telepathy might well be summed up in this phrase.) Magical thought often does seem absurd when its role in creating relationships between man, society, and nature is ignored, or when it creates as much disharmony as it should ideally prevent. But this union of the self with images drawn from the world it contemplates is also found in the higher reaches of organized religions, whose lower reaches may show as much idiotic hankering after illusory powers as in any tribal society. The fact is that this union is a rare state, difficult to experience when the mind is occupied with logical thought and neurotic preoccupations: it seems that we can only begin to understand it when we lay aside our usual categories of body and mind, image and action and reason, to see in what place they can all be experienced as one.

Voodoo, shamanism, and yoga all claim that a certain stage in the process they embody brings with it various occult powers. So do all mystical and magical systems. These powers seem to be dangerous to wield, perhaps because they emerge at a time when the relations between what we call body, soul, and spirit are not yet fully established, the powers having to do with psychophysical processes that then remain outside the ambit of consciousness. By using them as though they were different from the consciousness of oneself, it seems that they can gain a quasi-independent existence: the last, unifying step has been missed.

Consciousness appears normally as a private possession, in the very shape of our own separative selves. But the more we enlarge our self-awareness and resolve conflicting attitudes within ourselves, the more we see that consciousness is not a thing, or an ego, but an experience of relationship between disparate things and processes. It may first arise as self-consciousness, in the interaction of the senses with the motor activity of the body; it is enlarged to include the outer world when we begin to know, to name things and to act on the basis of symbolic thought. There appears to be an ideal moment in the development of the various mental faculties when the function of words and symbols is properly experienced: the equivalence which is felt between the name, the namer and the thing named is known to be the fact of consciousness itself, and the unity then experienced is often accompanied by what we call magical thinking in one context, ESP in another. As societies become complicated and neurotic tension increases, the system of symbolic reference breaks down and its organic meaning is hidden under a pile of social purpose and rational explanation. As many people have pointed out, what we know by reason is peculiarly unsusceptible to being shared telepathically: doubtless because the habit of reasoning encourages us to think in terms of body and mind, self and other, and thus to cut the act of consciousness in two. All ritual activity seems to be an attempt to bridge the separation thus made, and points to the simple if quite 'unreasonable' assumption that when metaphors are taken literally, and symbols are experienced as being the very structure of one's consciousness, then one is saved from the results of having conceived mind as separate from body. Whenever dissociation annuls this divisive habit, consciousness is experienced as unitary.

We are here at a critical point. If it is the symbol which creates consciousness, does ESP emerge when two people share the same symbols, the same organic preoccupations, the same field of consciousness? Does like in fact influence like? The experience enshrined in voodoo and shamanism shows that we need not conclude that an image reflecting the complexity of relationships is one thing, what it reflects a second, and the consciousness of it a third. If consciousness is an experience of relationship, we must also agree that things are related whether we are conscious of them or not, and it may well be that consciousness is inherent in all such relationships. Our minds normally move in this world of relationship in order to understand and control it, as something separate, because we have divided mind from body, intention from attitude. If we relearn the basic unity of such formally distinct aspects of ourselves, the distinction we had simultaneously made between self and other also falls way. We are then conscious of things not as objects separate from us but as part of the act of consciousness itself, which can then move in the world of relationships by means of the very relationships it experiences.

Is it then possible that symbols, by containing the field of relationships and providing the ground of consciousness, are responsible for what we call ESP? When these relationships become conscious, then they can be acted upon both by the imagination and by the body: though what this unitary action consists of, no one who is not conscious will experience, and because he does not become what he experiences, will not be able to understand.

APPENDIX

A GUIDE TO THE EXPERIMENTAL EVIDENCE FOR ESP

John Beloff

It is to the learned journals that the serious student will mainly need to turn for evidence. The principal current English language journals devoted exclusively to parapsychology or psychical research are:

1. (a) *Proceedings of the S.P.R.* 1882–. Intermittent.
   (b) *Journal of the S.P.R.* 1884–. Quarterly.
   The Society for Psychical Research (S.P.R.) of London was the first society of its kind in the world.
2. *Journal of the Amer. S.P.R.* 1907–. Quarterly.
   The American Society for Psychical Research (A.S.P.R.) of New York was founded in 1885 as an offshoot of the London Society. After the Second World War it coalesced with the now defunct Boston S.P.R. It also publishes intermittent Proceedings.
3. *Journal of Parapsychology.* 1937–. Quarterly.
   This is primarily the organ of J. B. Rhine and his collaborators of the Duke University Parapsychology Laboratory of Durham, N. Carolina, which, since 1964, has become the F.R.N.M. (Foundation for Research on the Nature of Man) Institute for Parapsychology.
4. *International Journal of Parapsychology.* 1959–. Quarterly.
   The organ of the Parapsychology Foundation Inc. of New York, a grant-giving body founded in 1951 by Mrs Eileen Garrett who is its president, to promote parapsychological studies in all parts of the world.

Two useful bibliographical summaries are:

1. *Psychical Research: A Selective Guide to Publications in English.* S.P.R. Pamphlet, 1949. Revised edition, 1959. Price 1s.

2. *Bibliography of Parapsychology.* Compiled by G. Zorab. New York: Parapsychology Foundation, 1957.

This covers some of the Continental work that is often ignored in other English language texts. The author is Hon. Secretary of the Netherlands S.P.R.

The following books by important pioneers of this field have become classics of the literature:

1. R. Tischner. *Telepathy and Clairvoyance.* London: Kegan Paul, 1925. Translated from the German.

    An account of some astonishingly successful tests of telepathy and clairvoyance which Dr. Tischner carried out on a small number of sensitives and mediums in Munich.

2. R. Warcollier. *Mind to Mind.* New York: Creative Age Press, 1948; Paperback edition: Collier Books, 1963.

    Based on a lecture which Warcollier delivered at the Sorbonne, Paris, in 1946 entitled 'A Contribution to the Study of Mental Imagery Through Telepathic Drawing'. Warcollier was one of the foremost French pioneers. His book *La Télépathie* (1921) was later expanded and brought out in English as *Experimental Telepathy.* Boston: S.P.R., 1938.

3. Upton Sinclair. *Mental Radio.* Springfield, Ill.: C. C. Thomas 1930. Revised second printing, 1962.

    A detailed and illustrated account of the famous experiments which this well-known American novelist carried out with his gifted wife, Craig Sinclair, as subject. The method is similar to that used by Warcollier with drawings as the target-material. The book is graced with a brief foreword by Albert Einstein.

4. J. B. Rhine. *Extra-Sensory Perception.* Boston: Bruce Humphries 1934.

    The book which first popularized the term 'ESP'. Dr. Rhine is, by common consent, the father of modern parapsychology. He standardized the card-guessing methodology which allows for an exact statistical test of the significance of any given result. He was the first to pursue parapsychology within an academic setting. This volume was followed by

    idem. *New Frontiers of the Mind.* London: Faber, 1937, Penguins 1950.

    idem. with four co-authors. *Extra-Sensory Perception after 60 years.* New York: Holt, 1940.

    idem. *The Reach of the Mind.* New York: Sloane, 1947; London: Faber, 1948.

    The first book to discuss the evidence for PK (Psycho-Kinesis), i.e. the influence of the subject's wishing on the fall of dice.

APPENDIX: EVIDENCE FOR ESP

idem, with J. G. Pratt. *Parapsychology: Frontier Science of the Mind.* Springfield, Ill.: C. C. Thomas, 1948; Oxford: Blackwell.

This comes nearer than any other text to being a handbook for the practical research worker.

5. S. G. Soal and F. Bateman. *Modern Experiments in Telepathy.* London: Faber, 1954.

Soal was the principal exponent of the new parapsychology in Britain. The book presents a detailed account of his investigations with his two star subjects: Basil Shackleton and Mrs. Gloria Stewart. Many authorities regard it as being still the most impressive fund of evidence that we have for the reality of ESP.

Idem, and H. T. Bowden. *The Mind Readers.* London: Faber, 1959.

Inspired by the idea that children would prove better subjects than adults, the book presents an account of a lengthy investigation of the Jones Boys, the remarkable Welsh telepathic cousins. Although the scoring was at a higher level than in any other guessing experiment of modern times the boys could not score when physically isolated from one another. This led to various conjectures regarding possible means of normal communication so that the evidential status of these findings remains controversial. The whole investigation has been fiercely but somewhat unfairly criticized by Hansel (see below).

6. L. L. Vasiliev. *Experiments in Mental Suggestion.* Church Crookham, Hants.: I.S.M.I. publications, 1963. Translated from the Russian.

Until his death in 1966 Vasiliev was the leading exponent of parapsychology in the Soviet Union. This book is an account of experiments he carried out during the 1930s in which a hypnotized subject is either awakened from or put back into a trance by means of signals transmitted telepathically from a distance. In one striking series the sender was in Sebastopol and the subject was in Leningrad! These experiments have only recently become generally known since the thaw made it possible once again for Russians to concern themselves with the paranormal.

The following books can be recommended as useful introductions to the topic that will appeal to a wide audience:

1. Whately Carington. *Telepathy.* London: Methuen, 1945.

The author, himself an important pioneer, gives an account of his own experiments in Part I, Sects. 23–9.

2. D. J. West. *Psychical Research Today.* London: Duckworth, 1954; Revised for Penguins, 1962.

The author was for a long time Research Officer of the S.P.R.

3. Gardner Murphy. *The Challenge of Psychical Research: A Primer of Parapsychology*. New York: Harper & Bros., 1961.

The author, an eminent American psychologist, is a convinced believer in the importance of parapsychology for a proper understanding of the mind.

4. C. D. Broad. *Lectures on Psychical Research*. London: Routledge & Kegan Paul, 1962.

Especially Sect. A. 'Guessing Experiments' where this distinguished English philosopher undertakes a detailed examination of the work of S. G. Soal, J. G. Pratt, and G. N. Tyrrell.

5. R. H. Thouless. *Experimental Psychical Research*. Penguins (Original), 1963.

The author, an eminent British experimental psychologist, discusses the various experimental problems involved in parapsychology.

6. J. G. Pratt. *Parapsychology: An Insider's View of ESP*. New York: Doubleday, 1964.

Dr. Pratt, formerly Rhine's chief co-worker, is probably the most experienced experimenter still active in parapsychology. The chapter dealing with his experiments on homing pigeons is particularly worth reading.

7. C. E. M. Hansel. *ESP: A Scientific Evaluation*. New York: Charles Scribners' Sons, 1966.

Professor Hansel is a British experimental psychologist who is well known as a critic of parapsychology. This book is his valiant and, in places, astonishingly ingenious attempt to demolish the entire case for ESP. It is, of course, no more 'scientific' than any other book on this list but it is well worth reading to discover just how far an attitude of unswerving scepticism can still be adhered to in the teeth of the evidence.

International
Library of Philosophy
& Scientific Method

Editor: Ted Honderich
Advisory Editor: Bernard Williams

List of titles, page three

International
Library of Psychology
Philosophy &
Scientific Method

Editor: C K Ogden

List of titles, page six

ROUTLEDGE AND KEGAN PAUL LTD
68 Carter Lane London EC4

International Library of Philosophy and Scientific Method
(Demy 8vo)

Allen, R. E. (Ed.)
Studies in Plato's Metaphysics
Contributors: J. L. Ackrill, R. E. Allen, R. S. Bluck, H. F. Cherniss, F. M. Cornford, R. C. Cross, P. T. Geach, R. Hackforth, W. F. Hicken, A. C. Lloyd, G. R. Morrow, G. E. L. Owen, G. Ryle, W. G. Runciman, G. Vlastos
464 pp. 1965. 70s.

Armstrong, D. M.
Perception and the Physical World
208 pp. 1961. (2nd Impression 1963.) 25s.

Bambrough, Renford (Ed.)
New Essays on Plato and Aristotle
Contributors: J. L. Ackrill, G. E. M. Anscombe, Renford Bambrough, R. M. Hare, D. M. MacKinnon, G. E. L. Owen, G. Ryle, G. Vlastos
184 pp. 1965. 28s.

Barry, Brian
Political Argument
382 pp. 1965, 50s.

Bird, Graham
Kant's Theory of Knowledge:
An Outline of One Central Argument in the *Critique of Pure Reason*
220 pp. 1962. (2nd Impression 1965.) 28s.

Brentano, Franz
The True and the Evident
Edited and narrated by Professor R. Chisholm
218 pp. 1965, 40s.

Broad, C. D.
Lectures on Psychical Research
Incorporating the Perrott Lectures given in Cambridge University in 1959 and 1960
461 pp. 1962. 56s.

Crombie, I. M.
An Examination of Plato's Doctrine
I. Plato on Man and Society
408 pp. 1962. 42s.
II. Plato on Knowledge and Reality
583 pp. 1963. 63s.

Day, John Patrick
Inductive Probability
352 pp. 1961. 40s.

International Library of Philosophy and Scientific Method
(Demy 8vo)

Edel, Abraham
Method in Ethical Theory
379 pp. 1963. 32s.

Flew, Anthony
Hume's Philosophy of Belief
A Study of his First "Inquiry"
296 pp. 1961. 30s.

Goldman, Lucien
The Hidden God
A Study of Tragic Vision in the *Pensées* of Pascal and the Tragedies of Racine. Translated from the French by Philip Thody
424 pp. 1964. 70s.

Hamlyn, D. W.
Sensation and Perception
A History of the Philosophy of Perception
222 pp. 1961. (2nd Impression 1963.) 25s.

Kemp, J.
Reason, Action and Morality
216 pp. 1964. 30s.

Körner, Stephan
Experience and Theory
An Essay in the Philosophy of Science
272 pp. 1966. 45s.

Lazerowitz, Morris
Studies in Metaphilosophy
276 pp. 1964. 35s.

Merleau-Ponty, M.
Phenomenology of Perception
Translated from the French by Colin Smith
487 pp. 1962. (2nd Impression 1965.) 56s.

Montefiore, Alan, and Williams, Bernard
British Analytical Philosophy
352 pp. 1965. 45s.

Perelman, Chaim
The Idea of Justice and the Problem of Argument
Introduction by H. L. A. Hart. Translated from the French by John Petrie
224 pp. 1963. 28s.

Schlesinger, G.
Method in the Physical Sciences
148 pp. 1963. 21s.

International Library of Philosophy and Scientific Method
(Demy 8vo)

Sellars, W. F.
Science, Perception and Reality
374 pp. 1963. 50s.

Shwayder, D. S.
The Stratification of Behaviour
A System of Definitions Propounded and Defended
428 pp. 1965. 56s.

Smart, J. J. C.
Philosophy and Scientific Realism
168 pp. 1963. (2nd Impression 1965.) 25s.

Smythies, J. R. (Ed.)
Brain and Mind
Contributors: Lord Brain, John Beloff, C. J. Ducasse, Antony Flew, Hartwig Kuhlenbeck, D. M. MacKay, H. H. Price, Anthony Quinton and J. R. Smythies
288 pp. 1965. 40s.

Taylor, Charles
The Explanation of Behaviour
288 pp. 1964. (2nd Impression 1965.) 40s.

Wittgenstein, Ludwig
Tractatus Logico-Philosophicus
The German text of the *Logisch-Philosophische Abhandlung* with a new translation by D. F. Pears and B. F. McGuinness. Introduction by Bertrand Russell
188 pp. 1961. (2nd Impression 1963.) 21s.

Wright, Georg Henrik Von
Norm and Action
A Logical Enquiry. The Gifford Lectures
232 pp. 1963. (2nd Impression 1964.) 32s.

The Varieties of Goodness
The Gifford Lectures
236 pp. 1963. (2nd Impression 1965.) 28s.

Zinkernagel, Peter
Conditions for Description
Translated from the Danish by Olaf Lindum
272 pp. 1962. 37s. 6d.

International Library of Psychology, Philosophy, and Scientific Method
(Demy 8vo)

PHILOSOPHY

Anton, John Peter
Aristotle's Theory of Contrariety
276 pp. 1957. 25s.

Bentham, J.
The Theory of Fictions
Introduction by C. K. Ogden
214 pp. 1932. 30s.

Black, Max
The Nature of Mathematics
A Critical Survey
242 pp. 1933. (5th Impression 1965.) 28s.

Bluck, R. S.
Plato's Phaedo
A Translation with Introduction, Notes and Appendices
226 pp. 1955. 21s.

Broad, C. D.
Ethics and the History of Philosophy
Selected Essays
296 pp. 1952. 25s.

Scientific Thought
556 pp. 1923. (4th Impression 1952.) 40s.

Five Types of Ethical Theory
322 pp. 1930. (8th Impression 1962.) 30s.

The Mind and Its Place in Nature
694 pp. 1925. (7th Impression 1962.) 55s. See also **Lean, Martin.**

Buchler, Justus (Ed.)
The Philosophy of Peirce
Selected Writings
412 pp. 1940. (3rd Impression 1956.) 35s.

Burtt, E. A.
The Metaphysical Foundations of Modern Physical Science
A Historical and Critical Essay
364 pp. 2nd (revised) edition 1932. (5th Impression 1964.) 35s.

International Library of Psychology, Philosophy, and Scientific Method
(Demy 8vo)

Carnap, Rudolf
The Logical Syntax of Language
Translated from the German by Amethe Smeaton
376 pp. 1937. (6th Impression 1964.) 40s.

Chwistek, Leon
The Limits of Science
Outline of Logic and of the Methodology of the Exact Sciences
With Introduction and Appendix by Helen Charlotte Brodie
414 pp. 2nd edition 1949. 32s.

Cornford, F. M.
Plato's Theory of Knowledge
The Theaetetus and Sophist of Plato
Translated with a running commentary
358 pp. 1935. (6th Impression 1964.) 28s.

Plato's Cosmology
The Timaeus of Plato
Translated with a running commentary
402 pp. Frontispiece. 1937. (4th Impression 1956.) 35s.

Plato and Parmenides
Parmenides' *Way of Truth* and Plato's *Parmenides*
Translated with a running commentary
280 pp 1939 (5th Impression 1964.) 32s.

Crawshay-Williams, Rupert
Methods and Criteria of Reasoning
An Inquiry into the Structure of Controversy
312 pp. 1957. 32s.

Fritz, Charles A.
Bertrand Russell's Construction of the External World
252 pp. 1952. 30s.

Hulme, T. E.
Speculations
Essays on Humanism and the Philosophy of Art
Edited by Herbert Read. Foreword and Frontispiece by Jacob Epstein
296 pp. 2nd edition 1936. (6th Impression 1965.) 32s.

Lange, Frederick Albert
The History of Materialism
And Criticism of its Present Importance
With an Introduction by Bertrand Russell, F.R.S. Translated from the German by Ernest Chester Thomas
1,146 pp. 1925. (3rd Impression 1957.) 70s.

International Library of Psychology, Philosophy, and Scientific Method
(Demy 8vo)

Lazerowitz, Morris
The Structure of Metaphysics
With a Foreword by John Wisdom
262 pp. 1955. (2nd Impression 1963.) 30s.

Lean, Martin
Sense-Perception and Matter
A Critical Analysis of C. D. Broad's Theory of Perception
234 pp. 1953. 25s.

Lodge, Rupert C.
Plato's Theory of Art
332 pp. 1953. 25s.
The Philosophy of Plato
366 pp. 1956. 32s.

Mannheim, Karl
Ideology and Utopia
An Introduction to the Sociology of Knowledge
With a Preface by Louis Wirth. Translated from the German by Louis Wirth and Edward Shils
360 pp. 1954. 28s.

Moore, G. E.
Philosophical Studies
360 pp. 1922. (6th Impression 1965.) 35s. See also **Ramsey, F. P.**

Ogden, C. K., and Richards, I. A.
The Meaning of Meaning
A Study of the Influence of Language upon Thought and of the Science of Symbolism
With supplementary essays by B. Malinowski and F. G. Crookshank.
394 pp. 10th Edition 1949. (4th Impression 1956) 32s.
See also **Bentham, J.**

Peirce, Charles, *see* **Buchler, J.**

Ramsey, Frank Plumpton
The Foundations of Mathematics and other Logical Essays
Edited by R. B. Braithwaite. Preface by G. E. Moore
318 pp. 1931. (4th Impression 1965.) 35s.

Richards, I. A.
Principles of Literary Criticism
312 pp. 2nd edition. 1926. (16th Impression 1963.) 25s.
Mencius on the Mind. Experiments in Multiple Definition
190 pp. 1932. (2nd Impression 1964.) 28s.

Russell, Bertrand, *see* **Fritz, C. A.; Lange, F. A.; Wittgenstein, L.**

International Library of Psychology, Philosophy, and Scientific Method
(Demy 8vo)

Smart, Ninian
Reasons and Faiths
An Investigation of Religious Discourse, Christian and Non-Christian
230 pp. 1958. (2nd Impression 1965.) 28s.

Vaihinger, H.
The Philosophy of As If
A System of the Theoretical, Practical and Religious Fictions of Mankind
Translated by C. K. Ogden
428 pp. 2nd edition 1935. (4th Impression 1965.) 45s.

von Wright, Georg Henrik
Logical Studies
214 pp. 1957. 28s.

Wittgenstein, Ludwig
Tractatus Logico-Philosophicus
With an Introduction by Bertrand Russell, F.R.S., German text with an English translation en regard
216 pp. 1922. (9th Impression 1962.) 21s.
For the Pears-McGuinness translation—*see page 5*

Zeller, Eduard
Outlines of the History of Greek Philosophy
Revised by Dr. Wilhelm Nestle. Translated from the German by L. R. Palmer
248 pp. 13th (revised) edition 1931. (5th Impression 1963.) 28s.

PSYCHOLOGY

Adler, Alfred
The Practice and Theory of Individual Psychology
Translated by P. Radin
368 pp. 2nd (revised) edition 1929. (8th Impression 1964.) 30s.

Bühler, Charlotte
The Mental Development of the Child
Translated from the German by Oscar Oeser
180 pp. 3 plates, 19 figures. 1930 (3rd Impression 1949.) 12s. 6d.

Eng, Helga
The Psychology of Children's Drawings
From the First Stroke to the Coloured Drawing
240 pp. 8 colour plates. 139 figures. 2nd edition 1954. (2nd Impression 1959.) 25s.

Jung, C. G.
Psychological Types
or The Psychology of Individuation
Translated from the German and with a Preface by H. Godwin Baynes
696 pp. 1923. (12th Impression 1964.) 45s.

International Library of Psychology, Philosophy, and Scientific Method
(Demy 8vo)

Koffka, Kurt
The Growth of the Mind
An Introduction to Child-Psychology
Translated from the German by Robert Morris Ogden
456 pp. 16 figures. 2nd edition (revised) 1928. (6th Impression 1952.) 45s.
Principles of Gestalt Psychology
740 pp. 112 figures. 39 tables. 1935. (5th Impression 1962.) 60s.

Kohler, W.
The Mentality of Apes
With an Appendix on the Psychology of Chimpanzees
Translated from the German by Ella Winter
352 pp. 9 plates. 19 figures. 2nd edition (revised) 1927. (4th Impression 1956.) 25s.

Malinowski, Bronislaw
Crime and Custom in Savage Society
152 pp. 6 plates. 1926. (7th Impression 1961.) 18s.
Sex and Repression in Savage Society
290 pp. 1927. (4th Impression 1953.) 21s.
See also Ogden, C. K.

Markey, John F.
The Symbolic Process and Its Integration in Children
A Study in Social Psychology
212 pp. 1928. 14s.

Murphy, Gardner
An Historical Introduction to Modern Psychology
488 pp. 5th edition (revised) 1949. (5th Impression 1964.) 40s.

Paget, R.
Human Speech
Some Observations, Experiments, and Conclusions as to the Nature, Origin, Purpose and Possible Improvement of Human Speech
374 pp. 5 plates. 1930. (2nd Impression 1963.) 42s.

Petermann, Bruno
The Gestalt Theory and the Problem of Configuration
Translated from the German by Meyer Fortes
364 pp. 20 figures. 1932. (2nd Impression 1950.) 25s.

Piaget, Jean
The Language and Thought of the Child
Preface by E. Claparède. Translated from the French by Marjorie Gabain
220 pp. 3rd edition (revised and enlarged) 1959. (2nd Impression 1962.) 30s.

International Library of Psychology, Philosophy, and Scientific Method *(Demy 8vo)*

Piaget, Jean *(continued)*
Judgment and Reasoning in the Child
Translated from the French by Marjorie Warden
276 pp. 1928 (3rd Impression 1962.) 25s.

The Child's Conception of the World
Translated from the French by Joan and Andrew Tomlinson
408 pp. 1929. (4th Impression 1964.) 40s.

The Child's Conception of Physical Causality
Translated from the French by Marjorie Gabain
(3rd Impression 1965.) 30s.

The Moral Judgment of the Child
Translated from the French by Marjorie Gabain
438 pp. 1932. (4th Impression 1965.) 35s.

The Psychology of Intelligence
Translated from the French by Malcolm Piercy and D. E. Berlyne
198 pp. 1950. (4th Impression 1964.) 18s.

The Child's Conception of Number
Translated from the French by C. Gattegno and F. M. Hodgson
266 pp. 1952. (3rd Impression 1964.) 25s.

The Origin of Intelligence in the Child
Translated from the French by Margaret Cook
448 pp. 1953. 35s.

The Child's Conception of Geometry
In collaboration with Bärbel Inhelder and Alina Szeminska. Translated from the French by E. A. Lunzer
428 pp. 1960. 45s.

Piaget, Jean and Inhelder, Bärbel
The Child's Conception of Space
Translated from the French by F. J. Langdon and J. L. Lunzer
512 pp. 29 figures. 1956 (2nd Impression 1963.) 42s.

Roback, A. A.
The Psychology of Character
With a Survey of Personality in General
786 pp. 3rd edition (revised and enlarged 1952.) 50s.

Smythies, J. R.
Analysis of Perception
With a Preface by Sir Russell Brain, Bt.
162 pp. 1956. 21s.

van der Hoop, J. H.
Character and the Unconscious
A Critical Exposition of the Psychology of Freud and Jung
Translated from the German by Elizabeth Trevelyan
240 pp. 1923. (2nd Impression 1950.) 20s.